XII Brazilian Meeting on Simulational Physics 2013

Journal of Physics: Conference Series Volume 487

Paraiba, Brazil
5 – 10 August 2013

Editors:

Joao Antonio Plascak
Alexandre Rosas

ISBN: 978-1-63266-708-3
ISSN: 1742-6588

Printed from e-media with permission by:

Curran Associates, Inc.
57 Morehouse Lane
Red Hook, NY 12571

Some format issues inherent in the e-media version may also appear in this print version.

Printed by Curran Associates, Inc. (2014)

For permission requests, please contact the Institute of Physics
at the address below.

Institute of Physics
Dirac House, Temple Back
Bristol BS1 6BE UK

Phone: 44 1 17 929 7481
Fax: 44 1 17 920 0979

techtracking@iop.org

Additional copies of this publication are available from:

Curran Associates, Inc.
57 Morehouse Lane
Red Hook, NY 12571 USA
Phone: 845-758-0400
Fax: 845-758-2634
Email: curran@proceedings.com
Web: www.proceedings.com

TABLE OF CONTENTS

Journal of Physics: Conference Series **487** (2014) 011001 doi:10.1088/1742-6596/487/1/011001

Preface

This special issue includes invited and selected articles of the VIIth Brazilian Meeting on Simulational Physics (BMSP), held in João Pessoa, Paraíba, Brazil, from the 5th to 10th August, 2013. This is the seventh such meeting, and the first one to have contributed papers published in the Journal of Physics: Conference Series.

The previous meetings in the BMSP series took place in the mountains of Minas Gerais and in the region of the Brazilian Pantanal. Now, for the first time, the Meeting was held in the pleasant shores of João Pessoa, the capital of the Paraíba state.

The VIIth BMSP brought together more than 50 researchers from all over the world for a vibrant and productive period. As in the previous meetings, the talks and posters highlighted recent advances in applications, algorithms, and implementations of computer simulation methods for the study of condensed matter, materials, out of equilibrium, quantum and biologically motivated systems. We are sure that this meeting series will continue to provide a valuable venue for people working in simulational physics to exchange ideas and discuss the state of art of this always expanding field.

We are very glad to realize this special issue, and are most appreciative to the editors of the Journal of Physics: Conference Series for making this publication possible. We are grateful for the outstanding work of the João Pessoa team, for the financial support of the Brazilian agencies CNPq, CAPES, FAPESQ, and of the Federal Universities UFPB and UFMG. At last, but not least, we would like to acknowledge all of the authors of this special issue for their contributions.

João Antonio Plascak
Alexandre Rosas
Guest Editors

Peer review statement

All papers published in this volume of *Journal of Physics: Conference Series* have been peer reviewed through processes administered by the proceedings Editors. Reviews were conducted by expert referees to the professional and scientific standards expected of a proceedings journal published by IOP Publishing.

Exploring new frontiers in statistical physics with a new, parallel Wang-Landau framework

Thomas Vogel[1,2], Ying Wai Li[1,3], Thomas Wüst[4], and David P. Landau[1]

1 Center for Simulational Physics, The University of Georgia, Athens, GA 30602, USA
2 Current address: Theoretical Division, Los Alamos National Laboratory, Los Alamos, NM 87545, USA
3 Current address: National Center for Computational Sciences, Oak Ridge National Laboratory, Oak Ridge, TN 37831, USA
4 Swiss Federal Research Institute WSL, Zürcherstrasse 111, CH-8903 Birmensdorf, Switzerland

E-mail: thomasvogel@physast.uga.edu

Abstract. Combining traditional Wang–Landau sampling for multiple replica systems with an exchange of densities of states between replicas, we describe a general framework for simulations on massively parallel Petaflop supercomputers. The advantages and general applicability of the method for simulations of complex systems are demonstrated for the classical 2D Potts spin model featuring a strong first-order transition and the self-assembly of lipid bilayers in amphiphilic solutions in a continuous model.

1. Introduction

In recent years the goal of Petaflop computing has been achieved by replying on massively parallel systems. This development requires a new approach to the efficient utilization of computing resources; moreover, improvements in methodology in computational statistical physics have also taken place. One such improvement, known as "Wang-Landau sampling", was described at the IIIrd Brazilian Meeting on Simulational Physics in 2003. In Wang–Landau (WL) sampling, the *a priori* unknown density of states $g(E)$ of a system is determined iteratively by performing a random walk in energy space (E) and sampling configurations with probability $1/g(E)$ (i.e. with a "flat histogram") [1, 2, 3, 4]. This procedure has proven very powerful for studying wide ranging problems with complex free energy landscapes because it circumvents the long time scales typically encountered near phase transitions or at low temperatures. The method also facilitates the calculation of thermodynamic quantities, including the free energy, at any temperature from a *single* simulation. Wang–Landau sampling is also a generic Monte Carlo procedure with only a few adjustable parameters, and it has been applied successfully to quite diverse problems including spin glasses, polymers, protein folding, lattice gauge theory, etc., see [5, 6, 7, 8] for examples. Over the years, the method itself was subject to multiple studies and various improvements to it have been proposed [9, 10, 11, 12, 13, 14, 15], but the underlying simplicity remains intact.

A few simple attempts at parallelization of the WL algorithm have been undertaken, but these are useful only for a relatively small number of processors (cores). One early approach

[2, 3] was to subdivide the total energy range into smaller sub-windows, each sampled by an independent WL random walker. The total simulation time is limited by the convergence of the slowest walker, but it can be tuned somewhat by an unequal partition of energy space. Nevertheless, an optimal load balancing is impossible due to the *a priori* unknown irregularities in the complex free energy landscape, and the individual energy intervals cannot be reduced arbitrarily due to systematic errors introduced because some regions of configurational space then become inaccessible.

An alternative scheme is to have multiple random walkers work simultaneously on the *same* density of states (and histogram). Although this approach seemingly avoids some problems, a recent, massively parallel implementation [16] revealed that correlations among the walkers could lead to a systematically underestimation of $g(E)$ in "difficult to access" energy regions. A proposed remedy to this problem was to add an "ad hoc" bias to the modification factor; but such inter-dependencies are highly problematic; besides the fact that an appropriate bias is, again, *a priori* unknown. It is also important to note that the effective round-trip times of the individual walkers are not improved by such an approach.

2. The Replica Exchange Wang–Landau algorithm

Our new approach [17] is a *generic* parallel method which combines the advantageous dynamics of the original Wang-Landau sampling scheme with the idea of replica-exchange Monte Carlo [18, 19]. We begin by splitting up the total energy range into h smaller sub-windows with large overlap between adjacent windows. The extent of the overlap o should be chosen to strike a balance between fast convergence of $g(E)$ and a reasonable exchange acceptance rate; an overlap of, e.g., $o \approx 75\%$ is a good choice [20]. See Fig. 1 for a visualization of a corresponding generic setup. Each energy sub-window is sampled by multiple (m), *independent* WL walkers. The key to this approach is that configurational or replica exchanges are allowed among WL instances of overlapping energy windows during the course of the simulation. As a consequence, each replica can travel through the entire energy space. The replica exchange move does not bias the overall WL procedure and, thus, guarantees the flexibility to apply it to any valid WL update/convergence rule (e.g. the $1/t$ algorithm [14]). Furthermore, our hierarchical parallelization approach imposes *no* principal limitation to the number of WL walkers or energy ranges so that it is possible to design setups which scale up to many thousands of CPUs.

The standard WL algorithm [1, 2] estimates the density of states, $g(E)$, in an energy range $[E_{\text{min}}, E_{\text{max}}]$ using a single random walker. During the simulation, trial moves are accepted with a probability $P = \min[1, g(E_{\text{old}})/g(E_{\text{new}})]$, where E_{old} (E_{new}) is the energy of the original (proposed) configuration. The estimation of $g(E)$ is continuously adjusted and improved using a modification factor f (as $g(E) \to f \times g(E)$) which progressively approaches unity as the

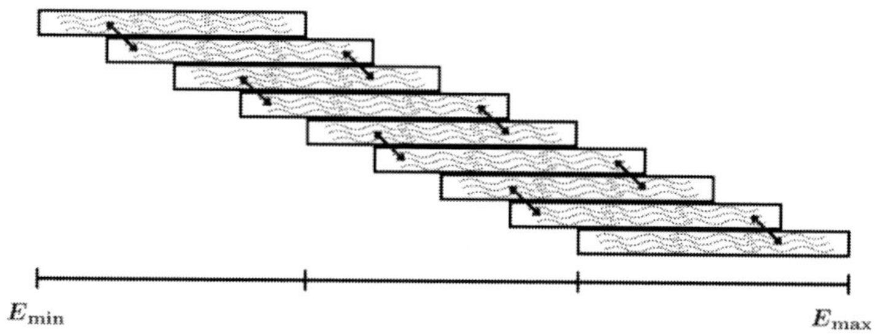

Figure 1. Example of a splitting of the global energy range into nine equal-size intervals with a constant overlap of 75%. There are nine walkers in each energy interval, arrows indicate the generic replica-exchange path.

simulation proceeds, while a histogram $H(E)$ keeps track of the number of visits to each energy E during a given iteration. When $H(E)$ is sufficiently "flat", the next iteration begins with $H(E)$ reset to zero and f reduced by some predefined rule (e.g. $f \to \sqrt{f}$). The simulation terminates when f reaches a small enough f_{final} at which point the accuracy of $g(E)$ is proportional to $\sqrt{f_{\text{final}}}$ for flat enough $H(E)$ [10].

In our parallel WL scheme, each random walker performs standard WL sampling in its energy sub-window. After a certain number of Monte Carlo steps, a "replica exchange" is proposed between two random walkers, i and j, where walker i chooses swap partner j from a neighboring sub-window at random. Let X and Y be the configurations that the random walkers i and j are carrying before the exchange; $E(X)$ and $E(Y)$ be their energies, respectively. From the detailed balance condition the acceptance probability P_{acc} for the exchange of configurations X and Y between walkers i and j is:

$$P_{acc} = \min\left[1, \frac{g_i(E(X))}{g_i(E(Y))}\frac{g_j(E(Y))}{g_j(E(X))}\right], \tag{1}$$

where $g_i(E(X))$ is the instantaneous estimator for the density of states of walker i at energy $E(X)$, cf. also [21].

An important new feature of our formalism is the provision of every random walker with its own $g(E)$ and $H(E)$ which are updated individually. Also, each walker must fulfill the WL flatness criterion *independently* at each iteration, ensuring that systematic errors as found in [16] cannot occur. When *all* random walkers within an energy sub-window have attained flat histograms, their estimators for $g(E)$ are averaged and redistributed among themselves before simultaneously proceeding to the next iteration. This procedure reduces the statistical error during the simulation with $\sqrt{m}$ [20], i.e. as for uncorrelated WL simulations. Furthermore, increasing m can improve the convergence of the WL sampling by reducing the risk of statistical outliers in $g(E)$ which would slow down subsequent iterations. Alternatively, it allows us, in principle, to use a weaker flatness criterion, leading to an additional potential speed-up [20].

The simulation terminates when all h energy intervals have attained f_{final}, and we are left with h fragments of $g(E)$ with overlapping energy intervals, see Fig. 2a for an example. The pieces are then used to calculate a single $g(E)$ over the entire energy range with the joining point for any two overlapping density of states pieces chosen where the inverse microcanonical temperatures $\beta = \mathrm{d}\log[g(E)]/\mathrm{d}E$ best coincide (see Fig. 2 b). This guarantees that there are no non-differentiable points in the microcanonical entropy $\log[g(E)]$, which would lead to artificial peaks in observables like the heat capacity. We set neighboring density-of-state pieces equal at those joining points and cut away superfluous parts to yield a continuous density of states, see Fig. 2 c. To estimate statistical errors, we repeat the simulation a few times and apply resampling methods (bootstrapping) [22]. Note that for each resampled global density of states, the joining points between neighboring pieces of data are at different positions.

3. How well does the algorithm work?

In order to assess the generality and performance of this novel parallel WL scheme, we applied it to multiple intrinsically different models. We shall concentrate on two examples here: The well studied 10-state Potts model in 2 dimensions showing a strong first-order transition, and a continuum model of amphiphilic molecules in solution showing multiple structural transitions during the lipid bilayer formation.

3.1. The 10-state Potts model and weak scaling

The Hamiltonian for the 10-state Potts model is given by

$$\mathcal{H} = -\sum_{\langle i,j\rangle} \delta(\sigma_i, \sigma_j), \tag{2}$$

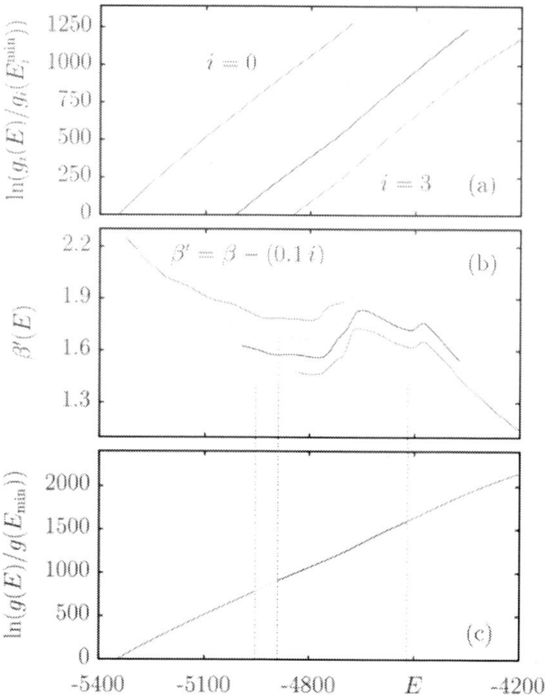

Figure 2. Illustration of the process of connecting density of states pieces after the simulation. (a) Individual pieces from four overlapping energy sub-windows (labeled i). (b) The inverse microcanonical temperature $\beta(E) = \mathrm{d}\ln[g(E)]/\mathrm{d}E$. Grid lines indicate the positions where neighboring data coincide best; individual curves are shifted with respect to each other for clarity. (c) Connected global density of states. The color codes show which part originates from which energy sub-window.

where the σ_i are spin variables which can take ten different (integer) values, and $\langle i, j\rangle$ refers to all nearest neighbor spin pairs. $N = L \times L$ is the total number of spins, L the linear lattice size.

Using single spin flip updates, serial generalized-ensemble Monte Carlo methods are able to study system sizes of about $N = 200^2$ within reasonable time (order of weeks) [2]. Figure 3 shows the density of states for the $N = 300^2$ lattice, which we obtained in a few hours by employing ≈ 2000 parallel walkers (the energy range is split into 717 sub-windows with an overlap of 75%, three walkers are deployed in each window), and the respective canonical distribution at the finite size transition temperature. The density of states covers almost 100 000 orders of magnitude and transition states at the phase transition are suppressed by a factor of $\approx 10^{-14}$. The finite-size transition temperature $T_c(L)$ is measured to be 0.70126. We estimate the critical temperature in the thermodynamic limit from results of system sizes up to 300^2 [20]. Our extrapolated value of $T_c^{\infty} = 0.701234 \pm 0.000006$ is in excellent agreement with the exactly known $T_c = 0.701232$ [23].

Besides this remarkable accuracy and absolute acceleration, our algorithm shows very good weak scaling behavior for these lattice models. This means that we can keep the simulation time almost constant when making the system larger if we also increase the number of computing cores by the same factor. In contrast, for serial runs this time increases rapidly with increasing system size. See Fig. 4 for illustration, where we show the simulation times for both serial and parallel runs. Absolute run times are rough estimates, h is the number of energy sub-windows for parallel runs.

3.2. Lipid bilayer formation in a continuous model

To show how this parallel approach can open up new vistas we also applied the method to a very distinct and particularly challenging molecular problem of high interest in biochemistry and pharmaceutical science: a coarse-grained continuum model for the self-assembly of amphiphilic molecules (lipids) in explicit solution. In this model, amphiphilic molecules, each of which is composed of a polar (P) head and two hydrophobic (H) tail monomers (P–H–H), are surrounded

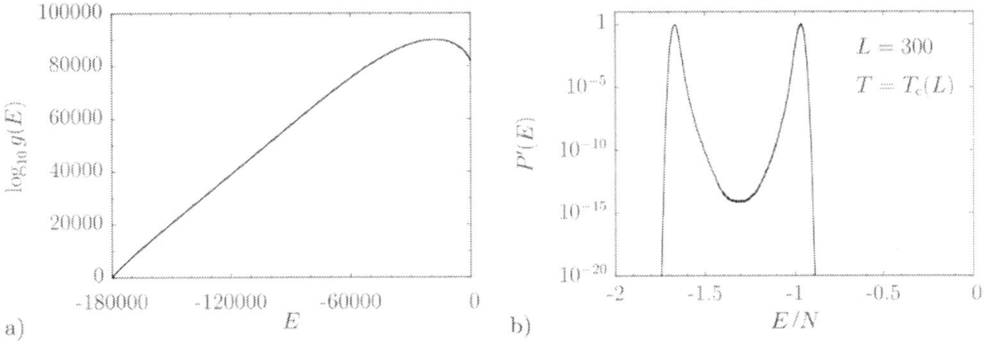

Figure 3. Data for the $L = 300$ 2D 10-state Potts model. (a) Logarithm of the density of states obtained by our parallel WL scheme with equal-size sub-windows. (b) Probability density (normalized to $P' = 1$ at the peaks) at the critical temperature. Transition states are suppressed by a factor of $\approx 10^{-14}$.

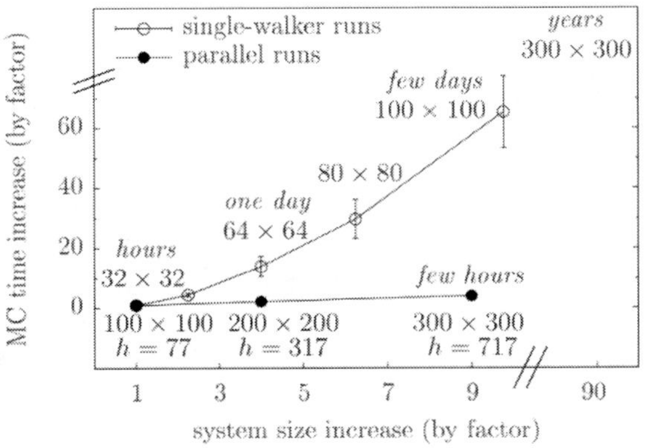

Figure 4. Relative MC time to terminate serial WL runs for different system sizes for the 2D Potts model (open circles) and for parallel runs (filled circles). The number of energy windows h increases according to the increase in system size.

by solvent particles (W), see Fig. 5 a. The interactions between H and W molecules as well as those between H and P molecules, are purely repulsive and given by the following potential:

$$U_{\rm rep}(r_{ij}) = 4\,\epsilon_{\rm rep}\left(\frac{\sigma_{\rm rep}}{r_{ij}}\right)^9 , \tag{3}$$

with r_{ij} being the distance between two particles. All other interactions between non-bonded particles are of Lennard-Jones type:

$$U_{\rm LJ}(r_{ij}) = 4\epsilon\left[\left(\frac{\sigma}{r_{ij}}\right)^{12} - \left(\frac{\sigma}{r_{ij}}\right)^{6}\right] . \tag{4}$$

As is usual when simulating generic coarse-grained models, we use reduced units, i.e., set $k_{\rm B} = 1$ and $\epsilon_{\rm rep} = \epsilon_{\rm X-Y} = 1$, where X-Y stands for H-H, P-P, P-W, and W-W interactions. Furthermore, we set $\sigma_{\rm rep} = 1.05\sigma$ with $\sigma = 1$, cf. [24, 25] for similar models. Bonded monomers are connected by a FENE-WCA [26, 27] potential:

$$U_{\rm FENE}(r_{ij}) = -0.5KR^2 \ln\left[1 - \left(\frac{r_{ij}}{R}\right)^2\right] + U_{\rm LJ}(r_{ij}) + \epsilon^* , \tag{5}$$

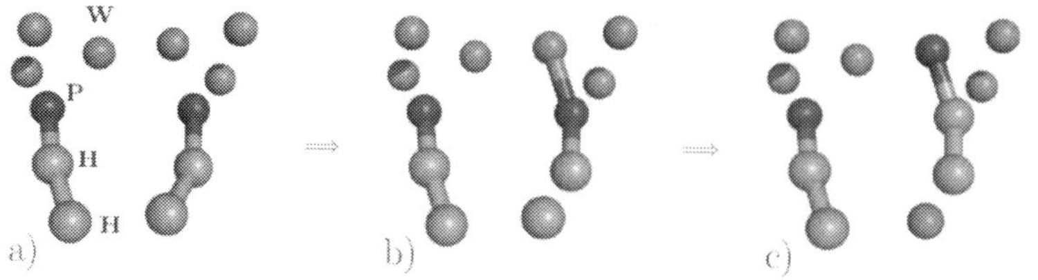

Figure 5. Reptation trial move: a) Amphiphilic molecules are composed of a polar (P) head bead and two hydrophobic (H) tail beads, and surrounded by solution particles (W). b) Intermediate step (not a valid configuration). c) Final reptation move configuration after particle reassignment.

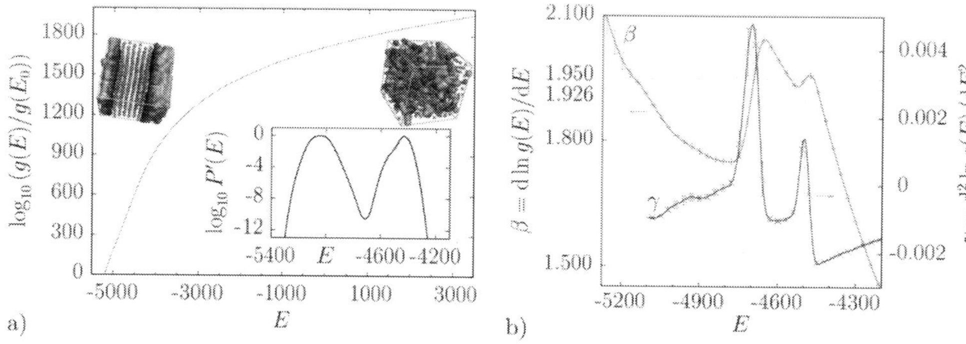

Figure 6. Data for the $N = 1000$-particle amphiphilic lipid model. (a) Logarithm of the density of states obtained with equal sub-windows as shown in Fig. 1 and normalized probability density at the critical temperature. Signs of a multi-stage transition can be seen in the right peak. Inset pictures show example configurations with $E \approx -5350$ and 3000, respectively. (b) First and second derivative of the microcanonical entropy ('microcanonical analysis'). See text for details.

where we set $K = 30$ and $R = 1.3$; and $\epsilon^* = \epsilon$ for $r_{ij} < 2^{1/6}\sigma$ or else 0.

Such models offer qualitatively different technical challenges compared to the Potts model and simulations of sufficiently large setups to study the bilayer formation are *impossible* for all practical purposes with traditional, single walker Monte Carlo methods. Besides local displacement moves, we introduce a reptation move which facilitates creation (and persistance) of highly ordered bilayer structures. During this move, a solution particle (W), randomly chosen amongst all which are within a distance R of either the head or the tail of a given amphiphilic molecule, is 'captured' and the respective other end of the lipid is 'released' (Fig. 5 b). To restore a valid configuration, particle types are then adapted correspondingly (Fig. 5 c). Note that due to generally different numbers of 'close' solution particles at both ends of the lipid, forward and backward moves have different selection probabilities, and this bias must be corrected in the Monte Carlo acceptance probability. We used a system containing $N = 1000$ particles in total, including $M = 125$ lipid molecules. One Monte Carlo sweep consists of N updates, on average $3M/10$ of which are reptation moves. The size of the simulation box is chosen such that the number density $\rho = 0.8$.

The density of states $g(E)$ over an energy range covering the complete lipid bilayer formation from random solutions to frozen bilayers spans ≈ 2000 orders of magnitude (Fig. 6 a) and cannot be sampled by single, serial Wang–Landau walkers in any reasonable time. During the bilayer formation multiple transitions occur, some almost simultaneously, as can be seen in the inset

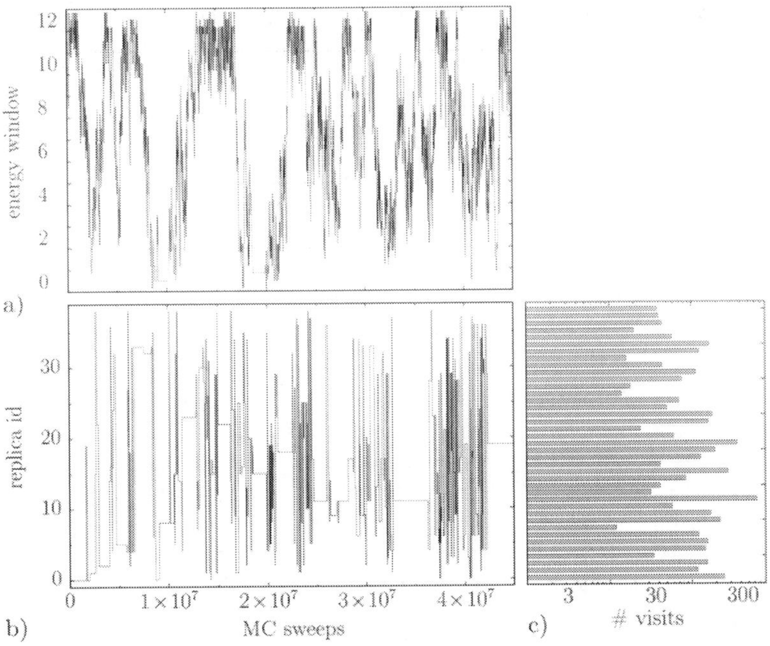

Figure 7. a) Path of a single replica (contributing to data shown in Fig. 6) through the 13 energy sub-windows during the first $\approx 4 \times 10^7$ MC sweeps. Replica exchanges are attempted every 10^4 sweeps. b) and c) Data from the viewpoint of one of the WL walkers in the lowest-energy sub-window. We plot the id of the replica which contributed to the histograms of that walker over time (b) and the corresponding histogram (c). All replicas contributed to the density of states estimator of that particular walker at some point (none of the histogram bins has zero entries).

of Fig. 6 a where a dip in the right peak of the canonical distribution is found. To understand such interlaced transitions we examine the microcanonical inverse temperature $\beta(E)$ and its derivative $\gamma(E)$ (see Fig. 6 b.) There is a clear, double back-bending in the inverse microcanonical temperature [28, 29], indicating first-order like transitions [30], and two well separated peaks in $\gamma(E)$. What we see here is a particular feature of the microcanonical point of view: The pre-transition occurs at higher energies, but lower (microcanonical) temperature. It is important to note that our algorithm does not miss such features.

Before we consider the physics of the lipid bilayer formation in detail, we need to comment on another technical aspect of our method. Due to the restricted local energy ranges, WL walkers are also restricted to the corresponding parts of the phase space with the potential of locking out parts of the overall phase space during the simulation. The replica-exchange mechanism serves as a countermeasure to this source of systematic error; hence, we must assure ourselves that each replica can actually walk through the entire energy space. As an example, in Fig. 7 a we show the path of a replica through energy sub-windows. In this setup, we split the energy range in 13 overlapping sub-windows and deploy three WL walkers in each window. At the beginning of the simulation one replica, with an energy inside the walker's energy sub-window, is assigned to each of the 39 walkers. Hence, from another point of view, one can demand that every WL walker, restricted to its small energy range, should be visited by all replicas during the simulation. This is shown in Fig. 7 (b and c) for a walker in the lowest-energy window.

The histogram of visits (Fig. 7 c) verifies that the walker is actually visited by all replica (no histogram bin is empty). Even though this analysis does not guarantee that we covered all of phase space during the simulation (in fact, one can never determine this), the opposite behavior could indicate systematic errors.

Returning to the physical results, we first separate the energy scale for the interaction between solution particles from all others [31], to get a more unaltered look at the actual transitions inside the lipid bilayer: we set $\epsilon_{W-W} = 0.7$. After estimating the density of states $g(E)$, we continued generating data with fixed weights $1/g(E)$ and measure distributions $H(E, O)$ for geometrical observables O to characterize the main structural phases. We measure the following observables:

The bond order parameter $S_{\text{mol}}^{\text{head}}$ *[32]:*

$$S_{\text{mol}}^{\text{head}} = 1/2\,\left(3\langle\cos^2\theta\rangle - 1\right) \tag{6}$$

is a measure for the average orientation of bonds with respect to a given direction in the simulation box. θ is the angle between a bond connecting the head and center bead in a lipid and the x, y, or z-direction, such that the order parameter is maximal [1], and $\langle\cdot\rangle$ denotes the average over the respective bonds of all lipids. For perfectly aligned bonds S_{mol} takes the maximum value of 1, its minimal value is -0.5, where all bonds are perpendicular to a certain axis, e.g., for lipids forming a perfect cylinder and taking the cylinder axis as reference. For spherical or random configurations the parameter is close to 0, independently of the reference axis.

Diagonal elements of the gyration tensor of the largest amphiphilic cluster and prolateness [33, 34]: To measure these quantities it is necessary to define and detect the largest cluster of amphiphilic molecules in the system at every measurement. This is efficiently done by a continuous implementation of the Hoshen-Kopelman algorithm [35]. In analogy to [25] two amphiphilic molecules are defined to be neighbors whenever a tail monomer of one molecule is within a distance $d < 1.4\sigma$ from a tail monomer of the other molecule. Furthermore, as the extension of clusters is usually larger than the simulation box size and in order to calculate meaningful quantities, all measurements are done in unfolded coordinates, i.e., all pieces of a cluster connected only via the periodic boundaries will be moved such that the whole cluster is connected in free space. The diagonal elements of the gyration tensor of the largest lipid cluster are calculated as follows:

$$Q_{ii} = \frac{1}{N}\sum_{j=1}^{N}\left(r_j^{(i)} - r_{\text{com}}^{(i)}\right), \tag{7}$$

where j labels an individual particle, r is the position vector, $i = 1, 2, 3$ indicates one of the principal directions x, y, or z and $r_{\text{com}}^{(i)}$ is the position of the center of mass of the unfolded cluster in that direction. We use N here for the total number of particles belonging to the cluster. We derive the prolateness P from the components of the gyration tensor in three dimensions:

$$P = 9\,\frac{\operatorname{Tr}\widetilde{Q}^3}{(\operatorname{Tr}Q)^3}, \tag{8}$$

where $\operatorname{Tr}Q$ is the trace of the gyration tensor and $\widetilde{Q} = Q - \lambda I$, with $\lambda = (\operatorname{Tr}Q)/3$ and I being the unity matrix. The prolateness approaches its maximal value of 2 for thin cylinders or rod like structures ($Q_{11} \gg Q_{22} \approx Q_{33}$) and its minimum value of $-1/4$ for disks ($Q_{11} \approx Q_{22} \gg Q_{33} \approx 0$).

[1] Here and in the following we will refrain from doing principal component analyses of the amphiphilic cluster, as they are computationally very expensive, and use the principal axes of the simulation box instead. However, it turns out that low-energy structures will mostly align along those directions.

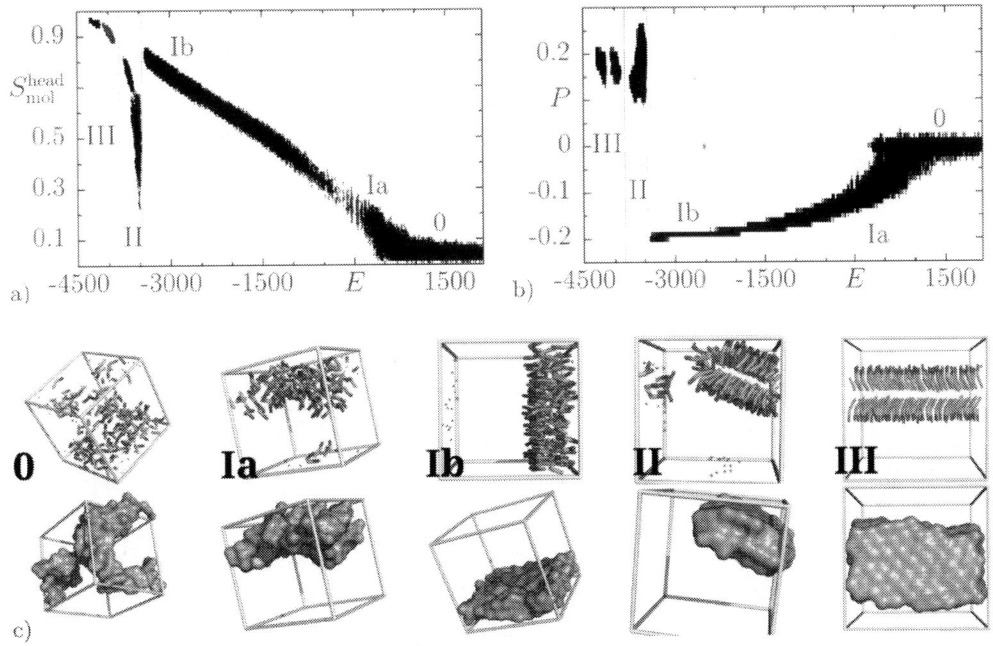

Figure 8. Analysis of main structural phases during the bilayer formation. Top views on the two-dimensional energy–bond-order (a) and energy-prolateness histograms (b) measured during a WL production run. c) shows representative structures. Top row: single lipid representation, bottom row: cluster representation. Solution particles (W) are not shown for clarity.

Figure 8 (a and b) shows top views of the histograms $H(E, S_{\rm mol}^{\rm head})$ and $H(E, P)$. Data is combined from all energy sub-windows, the grid-line positions correspond to energies at which $\gamma(E)$ (cf. Fig. 6) exhibits peaks. For $E \gtrsim 0$ we see disordered, random solutions ($S_{\rm mol}^{\rm head} \approx P \approx 0$; phase[2] '0'), at $E \approx 0$ cluster with a preferred orientation start to form ('Ia'). Those clusters become more and more disc-like ($P \to -0.25$) and lipid molecules align $S_{\rm mol}^{\rm head} \to 1$ ('Ib'), but are still 'liquid-like'. Phase 'II' can be seen as an intermediate phase where two lipid layers separate and finally form a bilayer with 'crystallized' lipids (phases 'III'). Figure 8 c shows representative structures from all phases with corresponding property sets ($E; S_{\rm mol}^{\rm head}; P$). For clarity, solution particles are not shown but were always included in simulations. For a comprehensive, more biochemical discussion of bilayer formation and structural changes in the bilayer, see [31].

Such a complete thermodynamic analysis of a system of this size, including explicit solution particles, was previously impossible using Monte Carlo techniques. The proposed parallel Wang–Landau method, however, enables us to enter a domain which had been dominated by molecular dynamics simulations, bringing along all the advantages of generalized ensemble MC techniques.

4. Summary

We have described a generic, parallel framework for Wang–Landau simulations based on the concepts of energy range splitting, replica exchange Monte Carlo, and multiple random walkers. The method is simple, generally applicable, and leads to significant advantages over traditional, single-walker WL sampling. Most importantly, it is readily adaptable to parallel systems with

[2] By 'phase' we refer to phases within this particular finite system.

an arbitrary number of processors (cores) including systems with many thousands of cores. As shown in two examples, our parallel WL framework allows us to successfully simulate previously inaccessible systems. The reasons are two-fold: first, each walker is now responsible for sampling a smaller configurational phase space, leading to faster convergence. Second, the replica exchange process revitalizes walkers in trapped states and avoids an erroneous bias in $g(E)$ due to potential ergodicity breaking. We demonstrated weak scaling behavior and the coverage of the entire energy space. Analyses of the strong scaling abilities and the influence of the number of walkers on the statistical error are given elsewhere [17, 20].

Acknowledgments

This work was supported by the National Science Foundation under Grants DMR-0810223 and OCI-0904685. Y.W. Li was partly sponsored by the Office of Advanced Scientific Computing Research; U.S. Department of Energy. Supercomputer time was provided by TACC under XSEDE grants PHY130009 and PHY130014. Assigned: LA-UR-13-26563.

References

[1] Wang F and Landau D P 2001 *Phys. Rev. Lett.* **86** 2050
[2] Wang F and Landau D P 2001 *Phys. Rev.* E **64** 056101
[3] Landau D P, Tsai S-H, and Exler M 2004 *Am. J. Phys.* **72** 1294
[4] Landau D P and Wang F 2003 *Braz. J. Phys.* **34** 354
[5] Rathore N and de Pablo J J 2002 *J. Chem. Phys.* **116** 7225
[6] Alder S, Trebst S, Hartmann A K, and Troyer M 2004 *J. Stat. Mech.* **2004** P07008
[7] Taylor M P, Paul W, and Binder K 2009 *J. Chem. Phys.* **131** 114907
[8] Langfeld K, Lucini B, and Rago A 2012 *Phys. Rev. Lett.* **109** 111601
[9] Yamaguchi C and Kawashima N 2002 *Phys. Rev.* E **65** 056710
[10] Zhou C and Bhatt R N 2005 *Phys. Rev.* E **72** 025701(R)
[11] Wu Y, Körner M, Colonna-Romano L, Trebst S, Gould H, Machta J, and Troyer M 2005 *Phys. Rev.* E **72** 046704
[12] Zhou C, Schulthess T C, Torbrügge S, and Landau D P 2006 *Phys. Rev. Lett.* **96** 120201
[13] Lee H K, Okabe Y, Landau D P 2006 *Comp. Phys. Comm.* **175** 36
[14] Belardinelli R and Pereyra V 2007 *Phys. Rev.* E **75** 046701
[15] Wüst T and Landau D P 2009 *Phys. Rev. Lett.* **102** 178101
[16] Yin J and Landau D P 2012 *Comp. Phys. Comm.* **183** 1568
[17] Vogel T, Li Y W, Wüst T, and Landau D P 2013 *Phys. Rev. Lett.* **110** 210603
[18] Geyer C J 1991 *Comp. Sci. Stat.: Proc. of the 23rd Symp. on the Interface* ed Keramidas E M (Fairfax Station, VA: Interface Foundation) p 156
[19] Hukushima K and Nemoto K 1996 *J. Phys. Soc. Jpn.* **65** 1604
[20] Vogel T, Li Y W, Wüst T, and Landau D P (in preparation)
[21] Nogawa T, Ito N, and Watanabe H 2011 *Phys. Rev.* E **84** 061107 (note the misprint in the corresponding equation though)
[22] Newman M E J and Barkema G T 1999 *Monte Carlo methods in statistical physics* (Oxford, New York: Oxford University Press)
[23] Baxter R J 1973 *J. Phys. C: Solid State Phys.* **6** L445
[24] Goetz R and Lipowsky R 1998 *J. Chem. Phys.* **108** 7397
[25] Fujiwara S, Itoh T, Hashimoto M, and Horiuchi R 2009 *J. Chem. Phys.* **130** 144901
[26] Warner H R 1972 *Ind. Eng. Chem. Fundam.* **11** 379
[27] Chandler D, Andersen H C 1971 *J. Chem. Phys.* **54** 26
[28] Gross D H E 2001 *Microcanonical Thermodynamics* (Singapore: World Scientific)
[29] Junghans C, Bachmann M, and Janke W 2006 *Phys. Rev. Lett.* **97** 218103
[30] Schnabel S, Seaton D T, Landau D P, and Bachmann M 2011 *Phys. Rev.* E **84** 011127
[31] Gai L, Vogel T, Maerzke K A, Iacovella C R, Landau D P, Cummings P T, and McCabe C 2013 *J. Chem. Phys.* **139** 054505
[32] Seelig A and Seelig J 1974 *Biochemistry* **13** 4839
[33] Aronovitz J A and Stephen M J 1987 *J. Phys. A: Math. Gen.* **20** 2539
[34] Blavatska V and Janke W 2010 *J. Chem. Phys.* **133** 184903
[35] Hoshen J and Kopelman R 1976 *Phys. Rev.* B **14** 3438

Simulating Phase Equilibria using Wang-Landau–Transition Matrix Monte Carlo

Katie A. Maerzke,[1] **Lili Gai,**[1] **Peter T. Cummings,**[1,2] **and Clare McCabe**[1,3,4]

[1] Department of Chemical and Biomolecular Engineering, Vanderbilt University, Nashville, TN 37235
[2] Center for Nanophase Materials Sciences, Oak Ridge National Laboratory, Oak Ridge, TN 37831
[3] Department of Chemistry, Vanderbilt University, Nashville, TN 37235

E-mail: c.mccabe@vanderbilt.edu

Abstract.
We will examine the strengths and weaknesses of the Wang-Landau and transition matrix Monte Carlo methods for simulating phase equilibria of continuous molecular systems alone and as a combined Wang-Landautransition matrix Monte Carlo algorithm. Although a combined Wang-Landau transition matrix Monte Carlo algorithm has been previously reported in the literature, the details of the method and a discussion of its performance for phase equilibria simulations has not been presented. The hybrid method combines the rapid initial estimate of the density of states from the Wang-Landau algorithm with the continual improvement in convergence of transition matrix Monte Carlo. The hybrid Wang-Landau–transition matrix (WL-TM) algorithm is found to be more efficient and has much better convergence properties than the Wang-Landau algorithm and is more robust than the transition matrix algorithm, enabling the simulations to reach relatively low reduced temperatures with ease.

1. Introduction

Computer simulations are becoming an increasingly useful tool for studying the structural and thermophysical properties of complex chemical systems. Simulation methods can be broadly classified into molecular dynamics (MD) and Monte Carlo (MC) approaches. Molecular dynamics simulations simply follow the system's trajectory through phase space by integrating Newton's equations of motion, whereas MC simulations generate configurations based on a prescribed probability distribution. Though the MD method is quite popular and used to study a wide variety of systems, using it to study phase equilibria can be quite challenging [1, 2, 3, 4, 5, 6]. However, a MC simulation does not have to obey any equations of motion; therefore, more advanced algorithms utilizing unphysical moves can be employed to accelerate the sampling of phase space.

A straightforward and conceptually simple MC method for simulating phase equilibria is the Gibbs ensemble [7, 8], which separates the coexisting phases into separate simulation boxes, eliminating the interface. Equilibrium is achieved through volume scaling moves and inter-box molecule transfers, which become challenging to perform for dense systems. With

[4] Corresponding author. Email: c.mccabe@vanderbilt.edu

the introduction of configurational-bias Monte Carlo [9] (CBMC), molecule insertions became feasible and the Gibbs ensemble became (and remains) widely used for simulations of phase equilibria. Grand canonical MC [10, 11] coupled with histogram reweighting [12, 13] and CBMC is another method for simulating phase equilibria which relies on the insertion and deletion of molecules. Though widely used, both the Gibbs and grand canonical ensembles have difficulty adequately sampling phase space for very dense systems and/or very low temperatures, even with the aid of CBMC.

Numerous other MC algorithms have been proposed to try to overcome the difficulties in sampling phase equilibria for dense fluids, including Gibbs-Duhem integration [14, 15], the NpT + test particle method [16, 17], and the broad histogram method [18, 19], along with a whole class of methods that force a broad sampling of phase space, such as expanded ensemble [20, 21], parallel tempering [22, 23], umbrella sampling [24], and multicanonical algorithms [25, 26, 27]. For an overview of these methods, the reader is referred to excellent review articles by Panagiotopoulos [28], Bruce and Wilding [29], and Orkoulas [30]. While not designed for phase equilibrium calculations, the Wang-Landau algorithm [31, 32] allows for the direct calculation of the density of states, or degeneracy, from which all thermodynamic properties can be computed, provided the relevant phase space is adequately sampled. As a result, phase equilibria can be calculated from a Wang-Landau simulation, typically at much higher cost than methods designed specifically for phase equilibria; on the other hand, many properties other than phase equilibria can be computed from the density of states determined in a Wang-Landau simulation. This and other density-of-states-based simulation methods have recently been reviewed by de Pablo et al. [33] and also feature in the book chapters by Shell et al. [34, 35]. The Wang-Landau algorithm belongs in the same class as the other methods, mentioned above, which achieve a broad sampling of phase space. These methods are also known as visited-states methods, since they keep track of how often the system visits each state and use the resulting data to extract thermodynamic properties. Visited states methods generally use an interative scheme to bias the system and achieve broad sampling of phase space, which can be inefficient. In contrast to visited-states methods, transition matrix methods [36, 37, 38, 39, 40, 41] track the statistics of the attempted transitions between states. Transition matrix methods do not employ an iterative biasing scheme, and thus never need to discard data, resulting in improved convergence properties.

In this paper we examine the Wang-Landau and transition matrix algorithms in the context of phase equilibria simulations for continuous molecular systems. Determining phase equilibria can be thought of as evaluating the relative free energies, or probabilities, of two coexisting phases [41]. We examine the strengths and weaknesses of each method, and show that combined they provide a more efficient and robust hybrid algorithm.

2. Methodology

2.1. Wang-Landau Algorithm

The Wang-Landau algorithm is a flat histogram method that enables us to directly determine the density of states, and thus all thermodynamic properties of interest [31, 32]. Simulations are generally carried out at constant density; thus, the density of states (DOS), degeneracy, or microcanonical ensemble partition function ($\Omega(N, V, E)$) is a function of the energy only ($\Omega(E)$). For continuous systems this requires dividing the energy into discrete intervals, or bins [42, 43]. In a standard Wang-Landau simulation, configurations are sampled with a probability proportional to the reciprocal of the density of states,

$$P(E) \propto \frac{1}{\Omega(E)} \tag{1}$$

Obviously, the DOS is unknown at the start of the simulation; therefore, the simulation begins with a value of $\Omega(E) = 1$ for every E in the specified energy range. Trial moves are accepted according to

$$\mathrm{acc}(i \to j) = \min\left\{1, \frac{\Omega(E_i)}{\Omega(E_j)}\right\} \tag{2}$$

After each move, the estimate of the DOS is updated with a modification factor, f

$$\ln\Omega(E) \to \ln\Omega(E) + \ln f \tag{3}$$

where the natural logarithm is used due to the many orders of magnitude difference between values of $\Omega(E)$. In addition to the density of states, a histogram (H) of visited states is also collected and incremented after each move

$$H(E) \to H(E) + 1 \tag{4}$$

The simulation continues until the histogram is sufficiently "flat," that is, until each value of $H(E)$ is not less than a specified percentage of the average value of $H(E)$,

$$H(E) \geq p \times \langle H(E) \rangle \tag{5}$$

where p is the flatness criterion. When the histogram is "flat," the modification factor is reduced according to $\ln f \to \ln f/2$, the histogram entries are reset to zero, and the simulation is continued. Though many authors have suggested variants of the modification factor update schedule [44, 45, 46, 47, 48, 49, 50, 51, 52, 53, 54, 55] the original modification factor update of simply dividing by two is still conventional. Once the modification factor becomes less than some predetermined value, the simulation ends. We should note that since the density of states is continuously updated, the Wang-Landau algorithm violates the condition of microscopic reversibility; however, by the end of the simulation the changes to the density of states are quite small and thus the condition is essentially satisfied.

A significant advantage of the Wang-Landau algorithm is that it allows one to very quickly determine a reasonable estimate of the density of states [56]. However, refining this estimate is very time-consuming, with the later iterations taking orders of magnitude longer to converge than the early stages and contributing relatively little to the overall DOS (due to the small modification factor). Moreover, the simulation eventually reaches a limiting accuracy, where further simulation does not yield more accurate results [57, 46, 48, 49, 50, 58, 53, 54]. Variations of the algorithm [46, 47, 48, 49, 58, 52, 53, 54], which primarily change the modification factor update schedule and/or histogram flatness criteria, have resulted in some improvements in accuracy and efficiency, with the $1/t$ algorithm of Belardinelli et al. showing particular promise. Both the original and modified versions of the WL algorithm have been primarily applied to spin systems and lattice models, though some have used the method to study continuous bead-spring and square-well polymers [59, 60, 61, 62, 63, 64], simple peptide models [65, 66, 67], methane dimers [58], and clusters of rigid water molecules [68]. The "standard" WL algorithm, by which we mean an algorithm that continuously refines an estimate of the density of states and samples configurations using the reciprocal of the DOS, has been used in a few instances to calculate the vapor-liquid coexistence curve for small systems of the Lennard-Jones fluid [43, 42]; however, this is a very expensive simulation since it requires the convergence of a two-dimensional density of states ($\Omega(N, E)$ or $\Omega(V, E)$).

More significant modifications to the original Wang-Landau algorithm have been proposed in an effort to obtain better performance in simulations of continuous molecular systems. Yan and de Pablo [57] developed two new methods which rely on either computing the configurational temperature, which requires calculating the forces, or a multimicrocanonical formalism. Though the methods do not exhibit the same saturation of the error as the conventional Wang-Landau algorithm, extension to simulations at non-constant density is not straightforward. Shell et al. introduced a hybrid Wang-Landau–transition matrix algorithm which relies on the relationship between the infinite temperature transition probabilities and the density of states [69]. This method has been used with mixed success by others to simulate simple peptides [70] and continuous spin models [71, 72]. Though the method improves the convergence for systems at constant density, extension to other ensembles requires a cumbersome four-dimensional matrix [34, 72]. The WL-TM method of Shell et al. will be discussed in more detail in Section 2.3.

The majority of Wang-Landau simulations are carried out at constant density; thus, the DOS is a function of only the energy. However, if we want to simulate phase equilibria, such as vapor-liquid coexistence, the DOS must be a function of the energy and density. Determining the two-dimensional, or joint, DOS requires significantly more simulation time and quickly becomes impractical, even for simple systems [43, 42, 73, 74, 75, 71, 76, 72]. Zhou et al. proposed using a "global update" (also known as "frontier sampling") to increase the efficiency of two-dimensional density of states simulations [77]; however, this method is very sensitive to the system-dependent parameterization and has been applied by others with only mixed success [47, 75, 78].

More significant modifications to the original Wang-Landau algorithm involve the inclusion of temperature to accelerate the convergence and increase the accuracy for continuous systems [79, 80]. Though it is often considered a strength of the WL algorithm that temperature is not explicitly included, we have found that for more complex molecular systems specifying the temperature is important for achieving accurate results [80]. Inclusion of advanced MC moves, such as configurational-bias [81], into an algorithm that determines the DOS as a function of the energy requires a re-derivation of the acceptance rule [80, 82, 83, 84], as configurations are no longer sampled according to their Boltzmann probability. For instance, the configurational-bias acceptance rule, which for Boltzmann sampling simulations is simply the ratio of the Rosenbluth weights, becomes the ratio of the Rosenbluth weights times the ratio of the density of states times an energetic factor $e^{\beta\Delta E}$ [80, 82, 83]. In the method of Ganzenmüller and Camp, the canonical ensemble partition function, $Q(N,V,T) = \sum_E e^{-\beta E}\Omega(N,V,E)$, is calculated for systems with fixed temperature and variable volume or number of molecules, reducing the calculation to a one-dimensional problem. Though this method has been used successfully to simulate vapor-liquid coexistence for a variety of molecules [85, 86, 87, 88, 89, 90, 91, 92, 93, 94, 95], its efficiency relative to other MC methods for determining phase equilibria is unclear, especially when the isobaric-isothermal ensemble is used. However, due to the use of a one-dimensional partition function rather than a two-dimensional density of states, the method is very likely much more efficient than a standard WL simulation. Others have developed hybrid MD/MC methods, where a short MD run is used to equilibrate the conformational degrees of freedom [96, 85]. Hybrid MD/Wang-Landau MC methods have been used to study protein folding [96, 97] and for isobaric-isothermal ensemble simulations of phase equilibria [85, 86, 87, 90, 91]. In this work, we consider a hybrid Wang-Landau algorithm by combining the best features of Wang-Landau with those of the transition matrix algorithm.

2.2. Transition Matrix Algorithm

In contrast to visited states methods, such as Wang-Landau, transition matrix (TM) Monte Carlo utilizes information about the *attempted transitions* between states to compute thermophysical properties in a variety of statistical mechanical ensembles [36, 37, 38, 39, 40, 41].

We will begin the derivation of the TM method with a brief overview of the standard Metropolis MC method. Each move in a Metropolis MC simulation consists of a two step process. First, a new microstate j is generated from the current microstate i, where a microstate is defined by the $\mathbf{r}^N$ coordinates of the N particles in the system. Next, the move to the new state j is accepted with probability $\mathrm{acc}(i \to j)$, given by the appropriate acceptance ratio:

$$\mathrm{acc}(i \to j) = \min\left\{1, \frac{p(j)}{p(i)}\right\} \tag{6}$$

where $p(i)$ is the probability of observing microstate i and depends on the choice of statistical mechanical ensemble. In the canonical (NVT) ensemble, $p(i)$ is given by

$$p(i) = \frac{1}{Q(N,V,\beta)} e^{-\beta E} \tag{7}$$

where Q is the partition function, E is the energy and $\beta = 1/k_{\mathrm{B}}T$, where k_{B} is Boltzmann's constant. In this case, Eqn. 6 reduces to the standard Metropolis acceptance rule

$$\mathrm{acc}(i \to j) = \min\left\{1, e^{-\beta \Delta E}\right\} \tag{8}$$

In the grand canonical (μVT) ensemble, $p(i)$ is given by

$$p(i) = \frac{1}{\Xi(\mu,V,\beta)} \frac{V^N}{\Lambda^{3N} N!} e^{-\beta E} e^{\beta \mu N} \tag{9}$$

where Ξ is the partition function, Λ is the thermal de Broglie wavelength, V is the volume, μ is the chemical potential, and N is the number of molecules. In addition to the temperature, the volume and chemical potential are fixed throughout the simulation; however, the number of molecules in the system changes through insertion and deletion moves. In the isobaric-isothermal (NpT) ensemble, $p(i)$ is given by

$$p(i) = \frac{1}{\Delta(N,p,\beta)} \frac{V^N}{\Lambda^{3N} N!} e^{-\beta E} e^{-\beta p V} \tag{10}$$

where Δ is the partition function. In this case, the temperature, number of molecules, and pressure are set and volume scaling moves are performed. Since the range of volumes may span several orders of magnitude, volume scaling moves are usually performed in the logarithm of the volume, rather than the volume itself. This changes the factor of V^N to V^{N+1} (see Frenkel and Smit [98] for more details). Note that for simplicity we are only considering unbiased moves in this discussion; however, biased moves can be performed using a suitable acceptance rule [98].

The goal of any simulation is to calculate macroscopic properties from a microscopic system. Thus we assign each microstate i to a macrostate I, which is defined by one or more macroscopic variables, such as energy (NVT), particle number (μVT), volume (NpT), or some other order parameter, depending on the ensemble and the properties of interest. Note that we use lower case letters to refer to microstates, and upper case letters for macrostates. The probability $P(I)$ is finding the system in macrostate I is given by the sum over all microstates i with the macrostate label I:

$$P(I) = \sum_{i \in I} p(i) \tag{11}$$

In a TM simulation, an extra step is performed after calculating the acceptance ratio. In this step, a collection matrix $C(I \to J)$ is updated with information about the attempted transition,

$$C(I \to J) = C(I \to J) + \mathrm{acc}(i \to j) \tag{12}$$

and

$$C(I \to I) = C(I \to I) + 1 - \mathrm{acc}(i \to j) \tag{13}$$

Journal of Physics: Conference Series **487** (2014) 012002 doi:10.1088/1742-6596/487/1/012002

regardless of whether or not the move was accepted.

The purpose of the collection matrix is to enable one to estimate the relative macrostate probabilities, P, which are related to the density of states, e.g., $P(N) \propto e^{\beta\mu N}Q(N,V,\beta) = e^{\beta\mu N}\sum_E e^{-\beta E}\Omega(N,V,E)$, but clearly depend on temperature, β. At any point in the simulation, the macrostate transition probabilities, $\Pi(I \to J)$, can be estimated from the collection matrix $C(I \to J)$ using

$$\Pi(I \to J) = \frac{C(I \to J)}{\sum_{\Delta I} C(I \to I + \Delta I)} \tag{14}$$

where $I + \Delta I$ is a new macrostate. After determining the macrostate transition probabilities $\Pi(I \to J)$ the condition of microscopic reversibility can be used to determine the macrostate probabilities

$$P(I)\Pi(I \to J) = P(J)\Pi(J \to I) \tag{15}$$

In general, solving this equation is an overspecified problem and a minimization technique must be implemented in order to find a solution, such as that used by Shell et al. in their infinite temperature WL-TM algorithm, where the macrovariable is E [69]. However, if we restrict ourselves to defining macrostates by only one macrovariable and only attempt transitions between neighboring macrostates, a unique solution can be found through sequential evaluation; e.g., if the macrovariable is N, then

$$\ln P(N+1) = \ln P(N) + \ln\left(\frac{\Pi(N \to N+1)}{\Pi(N+1 \to N)}\right) \tag{16}$$

For the TM method to be effective, the system must sample an adequate number of transitions between all allowed states. However, the system of interest often includes macrostates whose relative probabilites differ by several orders of magnitude; for instance, when calculating a vapor-liquid coexistence point, the macrostates corresponding to the vapor and liquid densities have a much higher probability than the macrostates which correspond to intermediate densities. Thus, sampling all of phase space in a reasonable amount of simulation time requires the introduction of a biasing function, $\eta(I)$, which is inversely proportional to the current estimate of the macrostate probabilities:

$$\eta(I) = \ln\left(\frac{1}{P(I)}\right) = -\ln P(I) \tag{17}$$

With the introduction of the biasing function, the acceptance rule must be adjusted

$$\mathrm{acc}(i \to j) = \min\left\{1, \frac{\exp[\eta(J)]p(j)}{\exp[\eta(I)]p(i)}\right\} = \min\left\{1, \frac{P(I)}{P(J)}\frac{p(j)}{p(i)}\right\} \tag{18}$$

However, the key to the TM method is that the *unbiased* acceptance ratio is still used to update the collection matrix, C. The C matrix, which does not contain any information about the biasing function, is used to determine the macrostate probabilities (using the condition of microscopic reversibility), from which we can calculate the properties of interest (see below). Thus, even though the actual move acceptance rates will depend on the choice of biasing function, the TM algorithm does not violate microscopic reversibility since only the unbiased acceptance ratios are used in data collection and analysis. Therefore, the weighting function can be periodically updated *without* discarding the previous data, resulting in a continual improvement in the calculation of the relative macrostate probabilities. [36, 37, 38, 39, 40, 41] Once we have accurately determined the macrostate probabilites, we can calculate the thermodynamic properties of interest in a straightforward manner using the histogram reweighting techniques outlined below in Section 2.4. For a more detailed derivation of the method the reader is referred to the paper by Errington [41].

Transition matrix simulations have been shown to be very efficient [36, 99, 100] and have been widely used to study phase equilibria [41, 101, 102, 103, 104, 105, 106, 107, 108, 109, 110] and other complex systems such as adsorption [111, 112] and wetting at interfaces [113, 114, 115, 116, 117, 118]. The continual improvement in the values for the relative macrostate probabilities is a strength of the TM algorithm; however, the performance of the algorithm depends strongly on the weighting function, η [100]. Although in certain cases one might be able to obtain a reasonable estimate of the weighting function, e.g., by reweighting and stepping down in temperature along the coexistence curve [119], in general this is either not possible or highly inefficient, and thus the weighting function must be slowly built up throughout the course of the simulation. For systems with very large differences in the relative probabilities of the macrostates, this can be a very slow process. Another advantage of the TM algorithm is that since it using standard Boltzmann sampling acceptance rules (modified by the ratio of the macrostate probabilities or weights), it does not distinguish between simple, unbiased moves, and advanced moves. This means that biased MC moves can be incorporated using their existing acceptance rules (e.g., by using the Rosenbluth weights for configurational-bias moves) [103], whereas the WL algorithm requires the re-derivation of the acceptance rule for advanced moves [80, 82, 83, 84].

2.3. Wang-Landau–Transition Matrix Algorithm

By combining the Wang-Landau and transition matrix algorithms, one can utilize their strengths while reducing their weaknesses. The fast initial estimate of the density of states of the Wang-Landau algorithm can be used to offset the slow building up of the weighting function in the transition matrix method, while the continuous improvement in convergence of the transition matrix algorithm offsets the limiting accuracy and slow convergence of the later stages of the Wang-Landau method. Although this idea has been used by Shell et al. to develop a hybrid Wang-Landau infinite temperature transition matrix method for systems at constant density [69] and has been mentioned in connection with grand canonical and expanded ensemble simulations [100, 120, 118], the details of the method and a discussion of its performance for phase equilibria simulations has not been presented.

A Wang-Landau–transition matrix (WL-TM) simulation begins with a Wang-Landau phase to obtain a reasonable estimate of the weights, η. This stage proceeds in the same manner as a standard Wang-Landau simulation, with an update of the macroscopic probabilities at every step,

$$\ln P(I) \rightarrow \ln P(I) + \ln f \tag{19}$$

where $\eta(I) = -\ln P(I)$, I is the macrovariable, f is the modification factor. Since we only need an *estimate* of the weights, we can often use very loose convergence criteria, e.g., $p = 30 - 40\%$ and final $\ln f = 10^{-3}$, which would be a very inappropriate choice for a standard Wang-Landau simulation [80]. However, we find that the even looser criteria of Shell et al. [43, 69], which stipulates that the modification factor can be updated after each histogram bin has been visited at least once, results in a significant loss of accuracy, as has been seen by others [71]. During the Wang-Landau phase, microscopic reversibility is violated, especially during the initial stages when the modification factor is large. Though this greatly accelerates sampling, these non-equilibrium transitions cannot be incorporated into the C matrix. Thus, the C matrix is not updated until the final iteration is reached, a more reasonable estimate of the density of states is obtained, the modification factor is relatively small, and microscopic reversibility is approximately satisfied. After the Wang-Landau phase has converged, the TM algorithm is then used to periodically update the weights using the data in the C matrix. The simulation continues for a specified number of steps (or cycles) until we deem that the macrostate probabilities have converged. The same reweighting techniques as in TM simulations can then be used to calculate the properties of interest, such as the coexistence densities (see Section 2.4).

Though very similar, the WL-TM algorithm presented here differs from the previous WL-TM method of Shell et al. [69] in several important respects. The prior method employed a special case of the TM algorithm; namely, the infinite temperature limit, which is equivalent to the broad histogram method of Oliveira et al. [18, 19]. By using the infinite temperature limit, the C matrix is updated with zeros and ones, rather than the acceptance ratio. Using the acceptance ratio has been shown to provide superior convergence properties [39]. However, the infinite temperature limit allows one to relate the transition probabilities to the density of states directly, which then allows for the calculation of thermodynamic properties as a continuous function of temperature over a wide range of temperatures (providing the relevant energies we adequately sampled). The disadvantage of this approach is that if we wish to perform simulations at non-constant density, we must use a very cumbersome and memory-intensive four-dimensional collection matrix [34, 72]. Moreover, because the density of states (or transition probability matrix) would be a function of both the energy and the density, acceptance rules for complex MC moves will need to be re-derived [80, 82, 83, 84]. More importantly, a two-dimensional density of states is much more difficult and time-consuming to converge [42, 57, 79, 75]. Though we anticipate that the TM part of the algorithm would help improve the efficiency compared to the "standard" WL algorithm, the simulations would still require a large amount of time to reach an acceptable level of accuracy.

Our version of the WL-TM algorithm is designed for studying systems at variable density and uses only a single macrovariable (with a two-dimensional C matrix). The TM portion of the algorithm follows the method presented by Errington for simulating phase equilibria [41], where the macrovariable is the number of molecules (N) or the volume (V), rather than the energy (E). By specifying a single macrovariable, we only require one-dimensional flat histogram, which results in a significant improvement in efficiency. A possible downside of eliminating the energy as the macrovariable is that the temperature must be specified, which means that multiple simulations must be run in order to compute the entire coexistence curve. However, we have found that for complex molecular systems including an explicit temperature results in increased accuracy in conformational sampling [80]; thus, this is not a significant drawback to our WL-TM method. Moreover, by using histogram reweighting techniques (see Section 2.4, below) the coexistence properties as a continuous function of temperature can still be obtained. We should also note that WL simulations, especially those with a two-dimensional density of states, typically employ multiple "windows" and thus also require multiple simulations; we note, however, that some authors have claimed to be able to obtain the complete density of states from a single simulation [42]. Moreover, because our WL-TM algorithm uses standard Boltzmann sampling acceptance rules (modified by the ratio of the macrostate probabilities or weights), rather than the ratio of the density of states, existing advanced MC moves can be used without modification, as is the case for the TM method [103], which greatly simplifies the implementation of the WL-TM algorithm in existing simulation codes.

2.4. Histogram Reweighting

Once the simulation has converged, we use standard histogram reweighting techniques [12, 13] to determine the coexistence properties. Histogram reweighting is often used in conjunction with simulations in the grand canonical ensemble; thus, the following discussion assumes we are working in that ensemble. Extension to the NpT ensemble should be straightforward.

Determining the coexistence properties at the specified simulation temperature requires adjusting the chemical potential while keeping the temperature fixed using the relationship

$$\ln P(N; \mu) = \ln P(N; \mu_0) + \beta(\mu - \mu_0)N \tag{20}$$

where μ_0 is the chemical potential used in the simulation. To determine the coexistence point, we specify the particle number N_{div} that divides the vapor and liquid phases, where N_{div} is the molecule number that corresponds to the minimum probability $P(N)$ between the liquid and vapor phases, and adjust the chemical potential until the area under the liquid and vapor curves is equal. In addition, if the zero-particle limit has been sampled, the vapor pressure can be calculated using

$$\beta pV = \ln\left[\sum_N P(N;\mu_{\text{coex}})\right] - \ln P(O;\mu_{\text{coex}}) - \ln 2 \tag{21}$$

where we have used the ideal gas as a reference state.

The calculation of coexistence densities over a range of temperatures requires some additional analysis. The goal is to determine the joint probability distribution $P(N,E)$, from which coexistence properties can be calculated. Rather than calculate the joint probability distribution directly, which would involve a four-dimensional transition matrix and a two-dimensional flat histogram, we follow the recommendation of Errington [41] and instead keep track of a histogram of visited states, $H(N,E)$, much like the Wang-Landau algorithm (though the TM histogram $H(N,E)$ is used for analysis purposes only). This requires a discretization of the energy; however, provided a reasonable energy interval is chosen the accuracy of the results does not depend strongly on the energy bin width. Alternatively, we could periodically record the molecule number and energy and extract the probability distribution at the end of the simulation or during the analysis, as is commonly done for multicomponent grand canonical simulations [121]. After normalizing the histogram according to

$$h(N,E) = \frac{H(N,E)}{\sum_E H(N,E)} \tag{22}$$

we can calculate the joint probability distribution using

$$P(N,E) = h(N,E)P(N) \tag{23}$$

where $P(N)$ is the probability of observing a macrostate with N molecules calculated using the TM algorithm (see Eqn. 16).

After calculating the joint probability distribution at temperature β, histogram reweighting [12] can be used to determine the coexistence properties at temperatures (β) and chemical potentials (μ) other than those used to perform the simulation. Specifically,

$$\ln P(N,E;\mu,\beta) = \ln P(N,E;\mu_0,\beta_o) - (\beta-\beta_o)E + (\beta\mu - \beta_0\mu_0)N \tag{24}$$

where again the subscript 0 indicates the original simulation conditions. Eqn. 24 is valid only over the range of energies and densities sampled in the simulation. To extend the range of the reweighting procedure, the results from R simulations over a range of temperatures with overlapping energies and densities are combined in a self-consistent manner to determine the composite probability, $\mathcal{P}$:

$$\mathcal{P}(N,E;\mu,\beta) = \frac{\sum_{i=1}^{R} h_i(N,E)P_i(N)\exp(-\beta E + \beta\mu N)}{\sum_{i=1}^{R}\exp(-\beta_i E + \beta_i\mu_i N - W_i)} \tag{25}$$

where the W_i (also known as weights) are constants related to the free energy of the simulations at β_i and μ_i [13]. The constants are determined using the relationship

$$\exp(W_i) = \sum_N\sum_E \mathcal{P}(N,E;\mu_i,\beta_i) \tag{26}$$

Given an initial guess for the constants W_i, Eqns. 25 and 26 are iterated until they converge (e.g., the difference between the old and new weights is less than some specified tolerance). Once we have determined $\mathcal{P}$, the properties of interest can be calculated using

$$\langle M \rangle_{\mu,\beta} = \sum_E \sum_N M\mathcal{P}(N, E; \mu, \beta) \tag{27}$$

where M is some mechanical property. For an overview of histogram reweighting in the grand canonical ensemble, the reader is referred to the review by Panagiotopoulos [28].

The key is that histogram reweighting allows one to use the WL-TM (or TM) method and calculate the desired properties as a continuous function of temperature. Obtaining results over a range of temperatures from a relatively small number of simulations is generally considered an advantage of the WL algorithm; however, by employing standard analysis techniques we can obtain the same results using the WL-TM and TM methods, though we may need to perform a larger number of simulations.

3. Simulation Details

Vapor-liquid coexistence curves (VLCCs) were calculated for a series of linear alkanes using the transferable potentials for phase equilibria – united atom (TraPPE–UA) force field [122], which employs pseudoatoms for all CH_x groups for increased computational efficiency. The TraPPE–UA force field contains terms for bond stretching, angle bending, dihedral rotations, and nonbonded interactions. The molecules studied ranged from single-site methane (a Lennard-Jones particle with diameter σ and characteristic energy ϵ adjusted to reproduce the experimental methane VLCC) to n-butane, which is the shortest alkanes to include a dihedral potential. Alkanes provide a relatively simple test case for studying the phase equilibria of molecular systems [123, 103], and thus allow us to more easily compare the accuracy and efficiency of different simulation methods [100].

To compare the accuracy and efficiency of the methods, we calculated the coexistence curves in the grand canonical ensemble using both TM and WL-TM and in the isobaric-isothermal ensemble using WL-TM. Above reduced temperatures of 0.6, the convergence criteria for the WL phase are set to very loose values of $p = 0.40$ and final $\ln f = 10^{-3}$, though at low temperatures stricter criteria are necessary. These values were determined by performing simulations at moderately high temperatures for methane with very loose convergence criteria, then increasing the strictness of the convergence parameters until satisfactory coexistence properties were obtained. At moderately high temperatures the accuracy and efficiency of the simulation do not depend strongly on the convergence criteria, which is an advantage of the WL-TM method in that less fine-tuning of parameters is required. The simulations were run at progressively lower temperatures until problems with convergence began to occur, at which point the value for p was increased and/or the final $\ln f$ decreased until satisfactory results were again obtained. The strictest convergence criteria used for the WL-TM simulations was $p = 0.85$ and final $\ln f = 10^{-4}$. After the simulation switches over to the TM method, the weights (or macrostate probabilities) are updated every 2.5×10^5 steps in the grand canonical ensemble (1 step = 1 move) or 1000 cycles in the isobaric-isothermal ensemble (1 cycle = N moves; N = the number of molecules). We also ran simulations for methane at select temperatures using a one-dimensional WL method (1D-WL). The 1D-WL method used here is essentially the WL-TM method without ever switching over to the TM phase, which is very similar to other modifed WL methods [79]. To obtain accurate macrostate probabilities with the 1D-WL, much stricter convergence criteria of $p = 0.80$ and final $\ln f = 10^{-6}$ must be used [80, 71]. For analysis purposes, histogram data is collected after the first twelve WL iterations (e.g., the point at which a WL-TM simulation would switch over to TM). The 1D-WL method is included here for comparison purposes only;

it is unlikely that 1D-WL would be used as the simulation algorithm if computing vapor-liquid equilibrium is the only property of interest.

Simulations in the grand canonical ensemble were run with a constant box length of 36 Å and maximum molecule numbers up to 778 for methane, 550 for ethane, 420 for propane, and 330 for n-butane. In addition, isobaric-isothermal simulations were performed for 550 methane molecules. These relatively large system sizes (compared to prior WL simulations [43, 42]) were chosen to enable a more rigorous test of the method. A spherical center-of-mass based cutoff of 14 Å with analytic tail corrections [124] was employed for the nonbonded Lennard-Jones interactions. The MC moves in the grand canonical ensemble consisted of 70% insertions and deletions using coupled-decoupled configurational-bias MC (CBMC) [81] with the usual choices for the cut-offs and number of trial sites [80]. The remaining moves consisted of either 30% molecular displacements (for methane and ethane) or 20% molecular displacements and 10% CBMC regrowths (for propane and butane). The simulations were run for at least 1.5×10^8 MC steps, with longer times for WL-TM simulations at low temperatures. The isobaric-isothermal simulations consisted of $1-3\%$ volume scaling moves, resulting in an average of approximately $5-15$ attempted volume moves per cycle with the remaining moves consisting of translations. Volume scaling moves make up a much smaller percentage of the total moves than the insertion and deletion moves in the grand canonical ensemble due to the inherent expense in re-calculating the total system energy after all the coordinates are scaled. The simulations were run for at least 7×10^5 MC cycles, with again longer simulation times required at lower temperatures (up to 3×10^6 cycles). Maximum displacements for translational and rotational moves were updated every 2.5×10^5 MC steps (μVT) or 1000 MC cycles (NpT) and optimized to achieve approximately a 50% acceptance rate. Simulations were run at 5 K temperature intervals along the coexistence curves with chemical potentials (μVT) and pressures (NpT) near the coexistence value and energy bin width of 200 K for the histogram $H(N, E)$ used in analysis. To explore the dependence of the performance on system size, we performed grand-canonical simulations for methane using a variety of system sizes, ranging from half to double the volume used for all other simulations. Four independent simulations were performed at each temperature, with an additional eight simulations (for a total of twelve) at select temperatures for methane and butane to enable a more accurate comparison of the efficiency and robustness of the methods. The reported errors are the standard deviations calculated from the independent simulations.

The critical temperature (T_c) and density (ρ_c) were determined using the scaling law [125]

$$\rho_{\text{liq}} - \rho_{\text{vap}} = B\,(T - T_c)^{\beta} \tag{28}$$

where $\beta = 0.325$ and the law of rectilinear diameters [126]

$$\tfrac{1}{2}\left(\rho_{\text{liq}} + \rho_{\text{vap}}\right) = \rho_c + A\,(T - T_c) \tag{29}$$

where ρ_{liq} and ρ_{vap} are the liquid and vapor densities, respectively. For the TM and WL-TM simulations, the ten highest temperatures were considered in the determination of the critical properties.

Multiple simulation "windows" are often used to improve the efficiency of the Wang-Landau and transition matrix algorithms, with each window covering a specified overlapping range of energy, molecule numbers, and/or volume [36, 31, 32, 43, 127, 113, 128, 101, 129, 130]. Once the simulations in each window have converged, the results can be patched together to obtain the complete density of states (WL) or macrostate probabilities (TM, WL-TM). The windows must be carefully constructed to ensure that phase space is adequately sampled and the system does not become trapped in a metastable state [43, 129, 64]. Although the TM algorithm can combine the results from multiple windows in an exact manner (by incorporating all the data

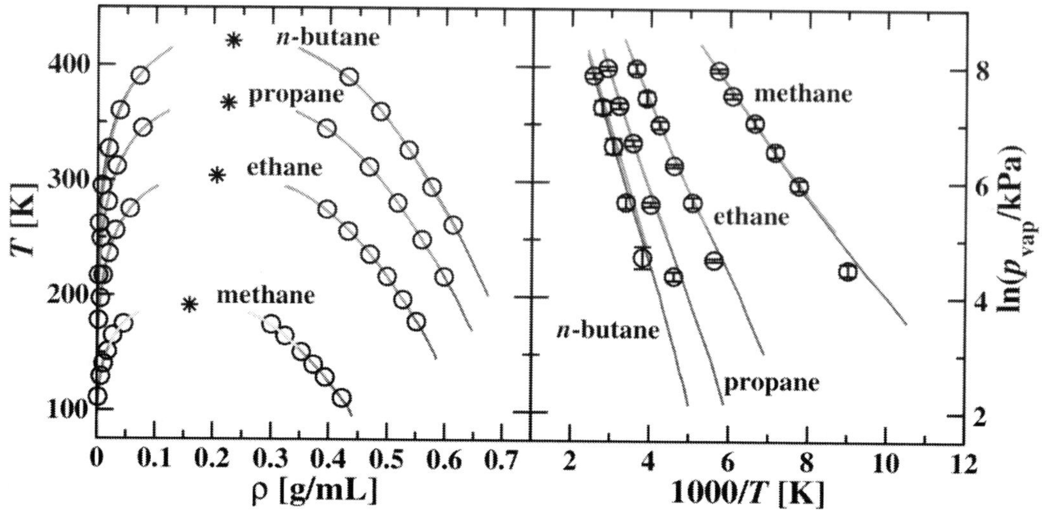

Figure 1. Vapor-liquid coexistence curves (left) and Clausius-Clapeyron plot (right) for methane, ethane, propane, and n-butane. GEMC (black circles) [122], grand canonical TM (red), grand canonical WL-TM (dark blue) and isobaric-isothermal WL-TM (light blue).

into a master transition matrix [36, 113]), the best way to combine the results from a WL simulation is not clear, especially when working with a two-dimensional density of states [131]. This is another potential advantage of the TM algorithm. Though utilizing multiple windows may be highly beneficial, especially in cases where efficiency is of great concern, we have not employed this technique and instead use a single simulation at each temperature. This allows us to compare the basic algorithms without the additional complexity of determining an appropriate windowing scheme.

4. Results and Discussion

The vapor-liquid coexistence curves and the Clausius-Clapeyron plot for all molecules included in this study can be found in Figure 1 and the critical temperature and density are reported in Table 1. Clearly, each method yields the correct coexistence curves and vapor pressures, and the critical points agree quite well with the previously published Gibbs ensemble results [122]. In fact, the TM and WL-TM simulations yield higher precision results, since grand canonical simulations can sample states closer to the critical temperature than Gibbs ensemble simulations. Obtaining the correct coexistence curves and vapor pressures is obviously essential; however, we are also concerned with the efficiency of the respective methods. Figure 2 shows the mean unsigned percentage error in the saturated liquid and vapor densities as a function of MC step for grand canonical simulations at temperatures $T^* = 0.9$, 0.8, and 0.7, where $T^* = T/T_C$. Due to the nature of the analysis, we cannot reliably calculate the coexistence properties (and thus also the average error) until we begin to see separated liquid and vapor peaks in the particle number probability distribution, $P(N)$, which for WL-TM simulations roughly corresponds to the end of the WL phase. For this reason, the lines in Figure 2 do not start at step zero, and the initial error, though larger than the final error, is still relatively small (about 2%). Both the TM and WL-TM methods show an initial rapid decrease in the average error, followed by a

Table 1. A Comparison of the Critical Temperature [K] and Density [g/mL] for Different Methods.[a]

molecule	method	T_C	ρ_C	T_{triple}/T_C
methane	GEMC[b]	191.4_{13}	0.160_3	
	TM	192.1_1	0.1625_2	
	WL-TM	192.0_1	0.1622_2	
	NpT WL-TM	192.2_3	0.1623_6	
	Exp.[c]	190.67_3	0.162_3	0.476
ethane	GEMC	304_2	0.206_3	
	TM	302.1_2	0.210_1	
	WL-TM	302.2_1	0.2117_2	
	Exp.	305.3_3	0.207_{12}	0.298
propane	GEMC	367_1	0.225_2	
	TM	366.2_1	0.2286_3	
	WL-TM	366.1_1	0.2284_2	
	Exp.	369.9_2	0.225_{18}	0.230
n-butane	GEMC	421_2	0.233_4	
	TM	420.7_3	0.2336_8	
	WL-TM	421.2_6	0.2342_7	
	Exp.	425_1	0.228_2	0.317

[a] Subscripts indicate uncertainties in the final digit(s).
[b] GEMC data taken from Ref. [122], with the exception of propane and n-butane, where some data points were re-calculated, resulting in slightly different critical properties.
[c] Experimental data taken from the NIST Chemistry WebBook [133].

slower decrease in error as the simulation progresses. The length of the simulation will depend on the degree of accuracy required; relatively short simulations of only 5×10^7 MC steps yield reasonably accurate results, with only a small improvement in accuracy for longer simulation times. The 1D-WL method takes a longer time to reach two-phase coexistence (hence the green lines in Figure 2 begin at a larger MC step number) and yield higher average percent errors in the saturated liquid and vapor densities. In fact, the 1D-WL method require mores than more than 10 times as many steps as the WL-TM method to converge at $T^* = 0.9$ and more than 100 times as many steps at $T^* = 0.7$. We should also note that a 1D-WL simulation converges much faster than a WL simulation with a two-dimensional density of states; thus, WL-TM is potentially orders of magnitude more efficient than a "standard" WL simulation for phase equilibria, which would require the density of states to be a function of both the energy and density. Thus, even though we must perform simulations at multiple temperatures, the WL-TM method is much more efficient than the unmodified WL algorithm. Another measure of efficiency, known as the tunneling time, is the average number of MC steps between visiting the extreme macrostates (e.g., the highest and lowest values of N) [132, 56]. While a very useful measure of efficiency for flat histogram methods, the mechanism for calculating the tunneling time was not implemented in the current version of our simulation code. We also find that the saturated liquid density converges somewhat faster than the saturated vapor density, due to larger fluctuations in the vapor phase.

Examination of Figure 2 seems to indicate that the WL-TM method does not offer any time-saving advantages over TM. Similar convergence behavior was seen by Shell et al. for small systems of Lennard-Jones particles at constant density [69]. However, what the figure does not show is that at temperatures of 0.8 and 0.7 only seven and five of twelve independent TM

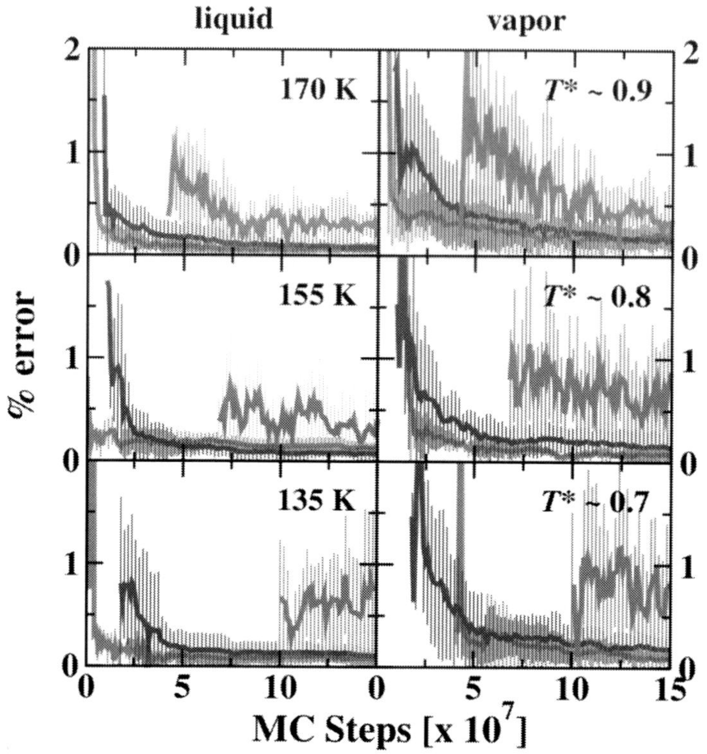

Figure 2. Mean unsigned percentage errors in the saturated liquid (left) and vapor densities (right) as a function of MC step in the grand canonical ensemble for methane using the TM (red), WL-TM (blue) and 1D-WL (green) methods at $T^* = 0.9$ (170 K, top), 0.8 (155 K, middle), and 0.7 (135 K, bottom). The lines begin at the point where we first see separated liquid and vapor peaks in the particle number probability distribution, which closely corresponds to the end of the Wang-Landau phase for WL-TM simulations. The target values are the average over twelve independent WL-TM simulations. Note that at $T^* = 0.8$ and 0.7 K only seven and five of twelve TM simulations converged, respectively.

simulations converged, respectively, whereas all of the WL-TM simulations converged without any problems. The reason for this behavior is that it is much easier for the simulation to become trapped at an intermediate state using the TM method, due to the slower build-up of the appropriate sampling weights. This problem is less severe at higher temperatures, where the free energy barrier between the vapor and liquid phases is relatively small (see Figure 3). In Figure 4 we see the same plot of average unsigned percentage error vs MC step for the n-butane simulations. The convergence behavior is the same for the somewhat larger molecule as for methane (a simple Lennard-Jones particle), indicating that the method performs well for more complex molecular systems. Again, from Figure 4 it appears that WL-TM and TM have the same rate of convergence, but the figure does not show that only approximately half of the TM simulations converge at $T^* = 0.8$ and 0.7, whereas all of the WL-TM simulations converge without difficulty. Clearly, the WL-TM method is more robust than the TM method.

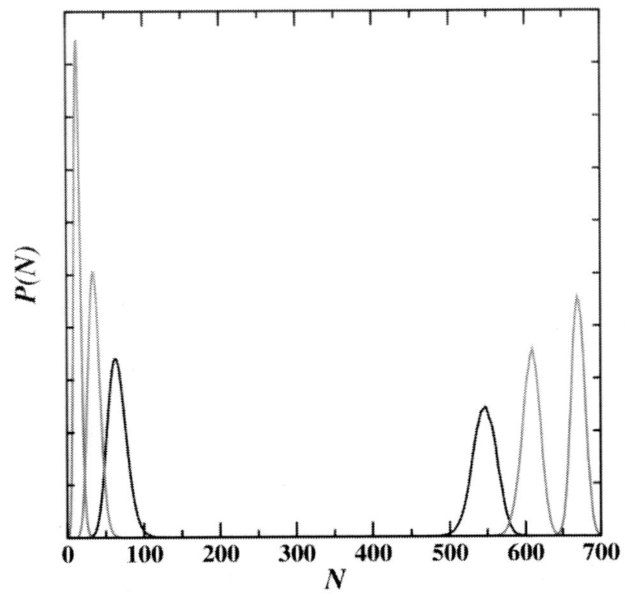

Figure 3. Coexistence particle number probabilities, $P(N)$, for methane at $T^* = 0.9$ (170 K, black), 0.8 (155 K, red), and 0.7 (135 K, green).

Transition matrix simulations at low temperatures are certainly possible; in fact, we have run TM simulations at temperatures as low as $T^* = 0.6$. However, they are very dependent on the initial simulation conditions and the updating scheme employed for the weights, beginning at temperatures below 0.8. For relatively small system sizes (approximately half the volume employed in this work), the TM algorithm can reach temperature as low as $T^* = 0.5$ [41], but convergence becomes more problematic as system size increases. We should note that simulations initialized in the liquid phase had a higher convergence rate than those initialized in the vapor phase. This is in contrast to the results of Errington [41], indicating a small difference in the implementation of the updating of the weights, though the method is essentially the same. The WL-TM algorithm is much less sensitive to such details. Though some of the independent simulations converged slightly faster than others, all of them converged in approximately the same amount of simulation time, regardless of the initial conditions. The largest difference in the convergence rate among the independent simulations was found at very low temperatures, where sampling becomes problematic even for the WL algorithm (see discussion below). Thus, although the WL-TM algorithm unfortunately does not offer any significant increase in speed compared to the TM algorithm (at least at relatively high temperatures), it is much more robust and requires less expert fine-tuning of the implementation details. Moreover, we must remember that the TM algorithm itself is a highly efficient method [36, 99, 100], so by extension, WL-TM is also very efficient.

The robustness of the WL-TM algorithm is a significant advantage of the method. WL-TM simulations can easily reach reduced temperatures as low as 0.6 with loose criteria of $p = 0.4$ and final $\ln f = 10^{-3}$. Below $T^* = 0.6$, simulations are possible with stricter WL convergence criteria. Reduced temperatures below 0.5 were reached with convergence criteria ranging from

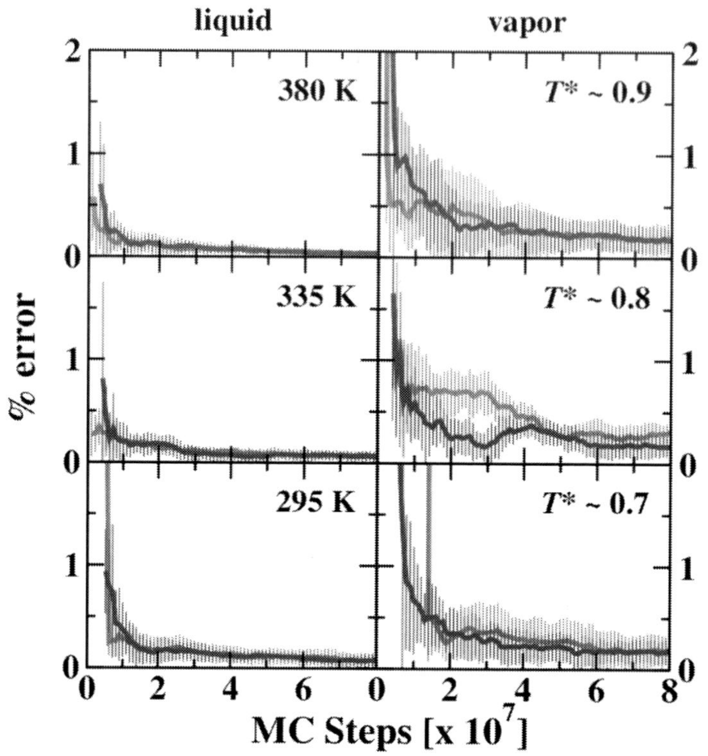

Figure 4. Mean unsigned percentage errors in the saturated liquid (left) and vapor densities (right) as a function of MC step in the grand canonical ensemble for n-butane using the TM (red) and WL-TM (blue) methods at $T^* = 0.9$ (380 K, top), 0.8 (335 K, middle), and 0.7 (295 K, bottom). The target values are the average over twelve independent WL-TM simulations. Note that at $T^* = 0.8$ and 0.7 only four and six of twelve TM simulations converged, respectively.

$p = 0.6 - 0.85$ and final $\ln f = 10^{-4}$. The strictest WL parameters were required for methane at 95 K ($T^* \approx 0.5$); although the experimental triple point for methane is approximately 91 K ($T^* = 0.476$), the triple point for the Lennard-Jones fluid is $T^* \approx 0.53$ [134, 135, 74, 136]. Thus, for our final simulation at $T^* = 0.5$ methane is in a metastable state. However, due to the large free energy barrier for nucleation, the system remains as a supercooled liquid well below the triple point, which is a well-known phenomenon for finite-size systems in cubic simulation boxes. However, at temperatures close to the triple point the simulation requires an inordinate amount of computer time, with the WL phase for methane requiring more than 100 times the number of MC steps at $T^* = 0.5$ as at 0.9 (10^7 vs. 10^9) and more than 40 hours on a 2.67 GHz Intel Xeon processor, followed by an additional period of TM simulations to refine the particle number probability distribution (up to 2×10^8 MC steps). Note that since the average error in the coexistence properties decreases rapidly after switching over from the WL to the TM phase and the time spent refining the macrostate probabilities in the TM phase depends on the degree of precision required, we have chosen to use the amount of time spent in the WL phase as an approximation for the length of time required to reach an acceptable degree of convergence. The experimental triple points for the other molecules are at lower reduced temperatures of 0.230 – 0.317 (see Table 1), which the TraPPE force field is able to reproduce reasonably well [137]. Thus, simulations at low reduced temperatures converge more easily for the longer alkanes than for methane, requiring 5 – 10 times fewer steps in the WL phase at $T^* = 0.5$. This is especially true for propane, which has the lowest triple point. However, even though the simulations require fewer numbers of MC steps than for methane ($\approx 10^8$ at $T^* = 0.5$), they still require a large amount of simulation time due to the increased number of interaction sites per molecule and the added complexity of the conformational sampling. For example, the WL phase for n-butane at $T^* = 0.5$ requires more than 48 hours of simulation time. At this point, the simulations begin to become impractical, and either a windowing technique to break the simulation into multiple parallel runs or possibly a more efficient modification factor update schedule for the WL phase might be beneficial.

We have also examined the dependence of the convergence rate on the system size by performing simulations for methane at temperature $T^* = 0.8$, 0.7, and 0.6 at volumes ranging from half to double the $(36\text{Å})^3 = 46656\text{Å}^3$ volume used in the rest of this work using the same convergence criteria of $p = 0.4$ and final $\ln f = 10^{-3}$ for the WL phase. All system sizes at the three temperatures studied yield the correct coexistence densities, with percentage errors of less than 1% (relative to the $V = 46656\text{Å}^3$ results). The exception is the smallest system at the lowest temperature, which overpredicts the saturated vapor density by approximately 5%. The vapor pressures are somewhat more sensitive the system size, with average percentage errors of $1 - 2\%$ at the two higher temperatures and 5% at the lowest temperature. The smallest system is again the exception, with an error of 11% in the vapor pressure. Clearly, one must take care to avoid finite size effects with very small systems. Unsurprisingly, we find that the simulations are faster for smaller systems and slower for larger systems. Figure 5 compares the number of MC steps spent in the WL phase for methane at the three different temperatures as a function of system size. We find that by doubling the volume, we increase the number of MC steps in the WL phase by a factor of $3 - 4$, whereas decreasing the volume by half reduces the number of MC steps by a factor of about 2.5. Moreover, the amount of wallclock time increases by a factor of 10 when the volume is doubled and decreases by a factor of 4 when the volume is halved. Thus, an appropriate system size should be large enough to avoid significant finite size effects, but small enough to avoid an unnecessary increase in simulation time.

In the NpT ensemble the method is much less efficient. As noted previously [41], TM simulations are very inefficient when volume is the macrovariable. The same is true for WL-TM. Since the volume scaling move requires that the total system energy be re-calculated, it is much more computationally expensive than an insertion or deletion move. However, to sample the volume probability distribution, we must perform a reasonably large number of volume moves. Thus, a balance must be struck between the more rapid convergence of the probability distribution and the slower performance of the simulation with an increased number of volume moves. Although it is difficult to relate grand canonical steps to isobaric-isothermal cycles, the amount of simulation time spent in the WL phase can be easily compared. At $T^* = 0.9$, the methane system takes 10^7 MC steps in the WL phase in the grand canonical ensemble, which takes approximately 15 minutes on a 2.67 GHz Intel Xeon processor. In contrast, in the isobaric-isothermal ensemble the system spends 7.5×10^5 cycles in the WL phase, which takes approximately 600 minutes on the same processor. As the temperature is decreased, the number of cycles in the WL phase increases, reaching 1.5×10^6 at 160 K ($T^* \approx 0.83$), which requires 20 hours of simulation time. Moreover, the WL phase is followed by at least the same number of cycles in the TM phase to refine the volume probability distribution. Given the inefficiency of the NpT ensemble for WL-TM simulations, we did not pursue lower temperatures or simulate the coexistence curves for other molecules. However, even though the method is impractical

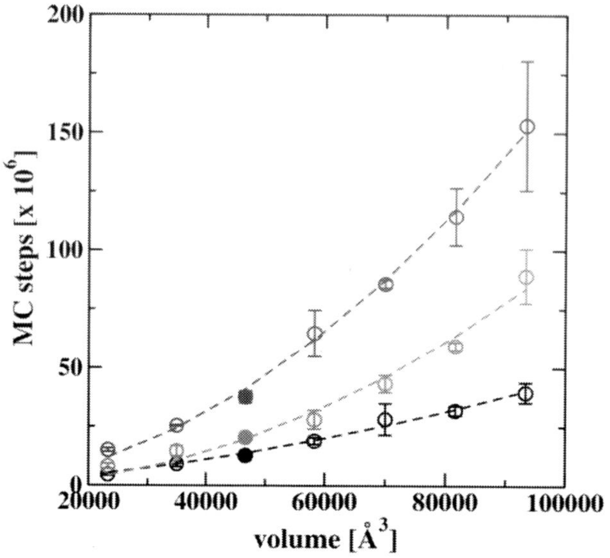

Figure 5. Number of MC steps in the WL phase as a function of system size for methane at $T = 0.8$ (black), 0.7 (red), and 0.6 (blue). The filled symbols represent the volume of (36 Å)^3 = 46656 Å^3 used for all other systems in this work, and the dashed lines are a quadratic fit to the data.

for vapor-liquid equilibria, it might be more applicable in other situations, such as for studying self-assembly, where the grand canonical ensemble is not always appropriate.

The efficiency of the WL-TM method might be improved by modifying the WL phase of the simulation. Many variations of the WL algorithm exist, most of which change the modification factor update schedule or histogram flatness criterion [69, 46, 47, 48, 49, 58, 52, 53, 54]. These methods mainly focus on improving the efficiency and accuracy of the later stages of a WL simulation, where the modification factor is quite small, convergence is slow, and the simulation eventually reaches a limiting accuracy. In the WL-TM algorithm we are only concerned with the initial stages, since the simulation switches over to the TM algorithm after the first few WL iterations. Thus, we have not investigated any of these modifications, though some do show significant improvements in efficiency, even in the early stages of the simulation, at least for lattice and spin systems. At low temperatures the simulation does spend a significant amount of time in the WL phase; thus, it might be worthwhile to explore some of the modifications to the WL algorithm if low temperature states are of great interest.

Simulations at very low temperatures remain problematic for reasons other than an inefficient choice of WL parameters and windowing scheme. At densities between the bulk coexistence densities, the system can form a variety of different structures, including a spherical droplet, cylindrical droplet, slab, cylindrical bubble, and spherical bubble (see Figure 6) [138, 139, 140]. The number of different structural transitions that must occur increases as the temperature decreases and/or the system size increases. For small systems at high temperatures, the system may pass through only one intermediate state (the slab), or possibly even go directly from a homogeneous vapor to a bulk liquid [138, 139], which explains why WL-TM (and TM) simulations are easier to perform for small systems and at high temperatures. For larger systems

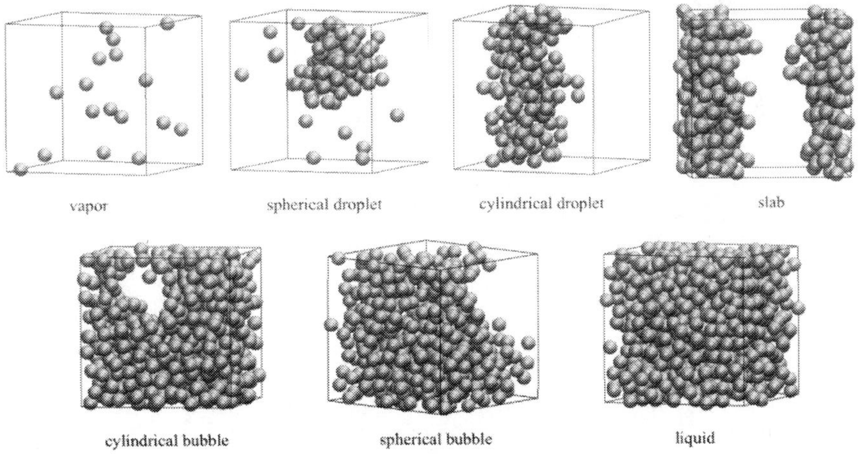

Figure 6. Simulation snapshots showing the structures found at varying densities, ranging from the vapor phase (top left) to the liquid phase (bottom right). The snapshots were taken from simulations of methane at $T^* = 0.7$ (135 K). Note the range of structures that must form at intermediate densities in order for the simulation to sample both coexisting phases.

at low temperatures, the system must pass through all the different structures to sample both the liquid and vapor phases. We should note that at these intermediate densities, the system is in an inhomogeneous state, with, for instance, a liquid-like droplet or cylinder surrounded by vapor. As the density changes, the shape of the condensed phase changes to minimize the surface area. The different shapes are due to the finite size and periodic boundaries used in our simulations. Since each structural change is associated with a barrier, [138, 139, 140] it becomes much easier for the system to become trapped in one of these intermediate stages. Thus, sampling becomes extraordinarily difficult, even when using the WL algorithm. Breaking the system into multiple windows becomes trickier; the system must be able to sample the different structural transitions in order for the resulting particle number probability distribution to be meaningful. Thus, deciding where to place the window divisions is difficult, and the windowing technique does not necessarily bring the large gain in efficiency that one might expect (since the system must still sample the structural transitions with large barriers). This problem is not limited to WL-TM; indeed, it is common to all methods that sample the states inside the region of phase coexistence. This problem can be avoided by simulating the liquid and vapor phases separately, as is done in standard grand canonical or Gibbs ensemble simulations. Of course, then we return to the problem of performing molecule insertions into dense phases. Multi-particle moves might alleviate the problem to a small extent; however, the structural rearrangements required to sample these transitions are so large that such moves would likely provide little benefit. A more promising approach is to perform temperature-expanded WL-TM simulations [117, 118, 20, 21], in which the liquid and vapor phases are sampled separately.

5. Conclusions

A hybrid Wang-Landau–transition matrix Monte Carlo method has been developed for efficient simulations of phase equilibria for continuous molecular systems. The method combines the strengths of both methods, using the Wang-Landau algorithm to obtain a rapid estimate for the density probability distribution (the transition matrix weights) and the transition matrix method to continually refine the probability distribution. The method is able to simulate phase equilibria with a one-dimensional probability distribution, resulting in a large increase in efficiency relative to the Wang-Landau algorithm, which requires a two-dimensional density of states to simulate systems at variable density, and also to a one-dimensional modified Wang-Landau algorithm. Incorporating the Wang-Landau algorithm makes the transition matrix method much more robust; that is, less sensitive to initial conditions and the fine details of the implementation. The Wang-Landau–transition matrix algorithm also enables the system to overcome larger free energy barriers, making simulations at lower temperatures for larger systems possible, with the caveat that as the temperature is lowered (or the system size is increased), the method, like all flat histogram methods, will eventually break down due to the large number of structural transitions inside the coexistence region.

Acknowledgements
Financial support from the National Science Foundation (#OCI-0904879) is gratefully acknowledged. A portion of the computational resources was provided by the National Energy Research Scientific Computing Center (NERSC). The authors would also like to thank David Landau and members of the Landau group at the University of Georgia for sharing their expertise with the Wang-Landau algorithm. K. A. M. would like to thank Jeff Potoff for his assistance in implementing grand canonical Monte Carlo and histogram reweighting, Ilja Siepmann for guidance with coupled-decoupled dual-cutoff CBMC, and Jeff Errington and Scott Shell for their repeated and gracious assistance in helping us to understand key aspects of their work.

[1] Harris J G 1992 *J. Phys. Chem.* **96** 5077–5086
[2] Alejandre J, Tildesley D J and Chapela G A 1995 *J. Chem. Phys.* **102** 4574–4583
[3] Mecke M, Winkelmann J and Fischer J 1997 *J. Chem. Phys.* **107** 9264–9270
[4] Gelb L D and Müller E A 2002 *Fluid Phase Equilib.* **203** 1–14
[5] Pamies J C, McCabe C, Cummings P T and Vega L F 2003 *Mol. Sim.* **29** 453–470
[6] Duque D and Vega L F 121 *J. Chem. Phys.* **121** 8611–8617
[7] Panagiotopoulos A Z 1987 *Mol. Phys.* **61** 813–826
[8] Panagiotopoulos A Z, Quirke N, Stapleton M and Tildesley D J 1988 *Mol. Phys.* **63** 527–545
[9] Siepmann J I 1990 *Mol. Phys.* **70** 1145–1158
[10] Norman G E and Filinov V S 1969 *High Temp.* **7** 216–222
[11] Adams D J 1974 *Mol. Phys.* **28** 1241–1252
[12] Ferrenberg A M and Swendsen R H 1988 *Phys. Rev. Lett.* **61** 2635–2638
[13] Ferrenberg A M and Swendsen R H 1989 *Phys. Rev. Lett.* **63** 1195–1198
[14] Kofke D A 1993 *Mol. Phys.* **78** 1331–1336
[15] Kofke D A 1993 *J. Chem. Phys.* **98** 4149–4162
[16] Möller D and Fischer J 1990 *Mol. Phys.* **69** 463–473
[17] Lotfi A, Vrabec J and Fischer J 1992 *Mol. Phys.* **76** 1319–1333
[18] de Oliveira P M C, Penna T J P and Herrmann H J 1998 *Eur. Phys. J. B* **1** 205–208
[19] de Oliveira P M C 1998 *Eur. Phys. J. B* **6** 111–115
[20] Lyubartsev A P, Martsinovski A A, Shevkunov S V and Vorontsov-Vel'yaminov P N 1992 *J. Chem. Phys.* **96** 1776–1783
[21] Escobedo F A and de Pablo J J 1995 *J. Chem. Phys.* **103** 2703–2710
[22] Marinari E and Parisi G 1992 *Europhys. Lett.* **19** 451–458
[23] Geyer C J and Thompson E A 1995 *J. Am. Stat. Soc.* **90** 909–920
[24] Torrie G M and Valleau J P 1977 *J. Comp. Phys.* **23** 187–199
[25] Berg B A and Neuhaus T 1991 *Phys. Lett. B* **267** 249–253
[26] Berg B A and Neuhaus T 1992 *Phys. Rev. Lett.* **68** 9–12
[27] Lee J 1993 *Phys. Rev. Lett.* **71** 211–214
[28] Panagiotopoulos A Z 2000 *J. Phys. Condens. Matter* **12** R25–R52
[29] Bruce A D and Wilding N B 2003 *Adv. Chem. Phys.* **127** 1–64
[30] Orkoulas G 2009 *Comput. Sci. Eng.* **64** 3668–3682
[31] Wang F and Landau D P 2001 *Phys. Rev. Lett.* **86** 2050–2053
[32] Wang F and Landau D P 2001 *Phys. Rev. E* **64** 056101
[33] Singh S, Chopra M and de Pablo J J 2012 *Annu. Rev. Chem. Biomol. Eng.* **3** 369–394
[34] Shell M S, Panagiotopoulos A Z and Pohorille A 2007 Methods Based on Probability Distributions and Histograms *Free Energy Calculations: Theory and Applications in Chemistry and Biology* ed Chipot C and Pohorille A (New York, NY: Springer) pp 77–118
[35] Shell M S and Panagiotopoulos A Z 2007 Methods for Examining Phase Equilibria *Free Energy Calculations: Theory and Applications in Chemistry and Biology* ed Chipot C and Pohorille A (New York, NY: Springer) pp 353–387
[36] Smith G R and Bruce A D 1995 *J. Phys. A: Math. Gen.* **28** 6623–6643
[37] Wang J S, Tay T K and Swendsen R H 1999 *Phys. Rev. Lett.* **82** 476–479
[38] Fitzgerald M, Picard R R and Silver R N 1999 *Europhys. Lett.* **46** 282–287
[39] Fitzgerald M, Picard R R and Silver R N 2000 *J. Stat. Phys.* **98** 321–345
[40] Wang J S and Swendsen R H 2002 *J. Stat. Phys.* **106** 245–285
[41] Errington J R 2003 *J. Chem. Phys.* **118** 9915–9925
[42] Yan Q, Faller R and de Pablo J J 2002 *J. Chem. Phys.* **116** 8745–8749
[43] Shell M S, Debendetti P G and Panagiotopoulos A Z 2002 *Phys. Rev. E* **66** 056703
[44] Zhou C and Bhatt R N 2005 *Phys. Rev. E* **72** 025701
[45] Jayasri D, Sastry V S S and Murthy K P N 2005 *Phys. Rev. E* **72** 036702
[46] Lee H K, Okabe Y and Landau D P 2006 *Comput. Phys. Commun.* **175** 36–40
[47] Poulain P, Calvo F, Antoine R, Broyer M and Dugourd P 2006 *Phys. Rev. E* **73** 056704
[48] Belardinelli R E and Pereyra V D 2007 *Phys. Rev. E* **75** 046701
[49] Belardinelli R E and Pereyra V D 2007 *J. Chem. Phys.* **127** 184105
[50] Belardinelli R E, Manzi S and Pereyra V D 2008 *Phys. Rev. E* **78** 067701
[51] Zhou C and Su J 2008 *Phys. Rev. E* **78** 046705
[52] Brown G, Odbadrakh K, Nicholson D M and Eisenbach M 2011 *Phys. Rev. E* **84** 065702
[53] Swetnam A D and Allen M P 2011 *J. Comput. Chem.* **32** 816–821
[54] Caparica A A and Cunha-Netto A G 2012 *Phys. Rev. E* **85** 046702

[55] Sinha S 2012 *Comput. Phys. Commun.* **183** 2616–2621
[56] Dayal P, Trebst S, Wessel S, Würtz D, Troyer M, Sabhapandit S and Coppersmith S N 2004 *Phys. Rev. Lett.* **92** 097201
[57] Yan Q and de Pablo J J 2003 *Phys. Rev. Lett.* **90** 035701
[58] Morozov A N and Lin S H 2009 *J. Chem. Phys.* **130** 215106
[59] Parsons D F and Williams D R M 2006 *Phys. Rev. E* **74** 041804
[60] Sliozberg Y and Abrams C F 2005 *Macromolecules* **38** 5321–5329
[61] Seaton D T, Mitchell S J and Landau D P 2008 *Braz. J. Phys.* **38** 48–53
[62] Seaton D T, Wüst T and Landau D P 2009 *Comput. Phys. Commun.* **180** 587–589
[63] Taylor M P, Paul W and Binder K 2009 *J. Chem. Phys.* **131** 114907
[64] Seaton D T, Wüst T and Landau D P 2010 *Phys. Rev. E* **81** 011802
[65] Gervais C, Wüst T, Landau D P and Xu Y 2009 *J. Chem. Phys.* **130** 215106
[66] Yang J S and Kwak W 2010 *Comput. Phys. Commun.* **181** 99–104
[67] Singh P, Sarkar S K and Bandyopadhyay P 2011 *Chem. Phys. Lett.* **514** 357–361
[68] Yin J and Landau D P 2011 *J. Chem. Phys.* **134** 074501
[69] Shell M S, Debenedetti P G and Panagiotopoulos A Z 2003 *J. Chem. Phys.* **119** 9406–9411
[70] Ghulghazaryan R G, Hayryan S and Hu C K 2006 *J. Comput. Chem.* **28** 715–726
[71] Bhar S and Roy S K 2009 *Comput. Phys. Comm.* **180** 699–707
[72] Bhar S and Roy S K 2013 *Comput. Phys. Comm.* **184** 1387–1394
[73] Mastny E A and de Pablo J J 2005 *J. Chem. Phys.* **122** 124109
[74] Mastny E A and de Pablo J J 2007 *J. Chem. Phys.* **127** 104504
[75] Mukhopadhyay K, Ghoshal N and Roy S K 2008 *Phys. Lett. A* **372** 3369–3374
[76] Volkov N A, Voronstsov-Velyaminov P N and Lyubartsev A P 2011 *Macromol. Theory Simul.* **20** 496–509
[77] Zhou C, Schulthess T C, Torbrügge S and Landau D P 2006 *Phys. Rev. Lett.* **96** 120201
[78] Yin J and Landau D P 2012 *Comput. Phys. Comm.* **183** 1568–1573
[79] Ganzenmüller G and Camp P J 2007 *J. Chem. Phys.* **127** 154504
[80] Maerzke K A, Gai L, Cummings P T and McCabe C 2012 *J. Chem. Phys.* **137** 204105
[81] Martin M G and Siepmann J I 1999 *J. Phys. Chem. B* **103** 4508–4517
[82] Siretskiy A, Elvingson C, Vorontsov-Velyaminov P and Khan M O 2011 *Phys. Rev. E* **84** 016702
[83] Radhakrishna M, Sharma S and Kumar S K 2012 *J. Chem. Phys.* **136** 114114
[84] Jain T S and de Pablo J J 2002 *J. Chem. Phys.* **116** 7238–7243
[85] Desgranges C and Delhommelle J 2009 *J. Chem. Phys.* **130** 244109
[86] Desgranges C, Hicks J M, Magness A and Delhommelle J 2010 *Mol. Phys.* **108** 151–158
[87] Aleksandrov T, Desgranges C and Delhommelle J 2012 *Fluid Phase Equilib.* **38** 1265–1270
[88] Ganzenmüller G and Camp P J 2011 *Condens. Matter Phys.* **14** 33602
[89] Schmidle H and Klapp S H L 2011 *J. Chem. Phys.* **134** 114903
[90] Aleksandrov T, Desgranges C and Delhommelle J 2012 *Mol. Sim.* **38** 1265–1270
[91] Desgranges C, Ngale K N and Delhommelle J 2012 *Fluid Phase Equilib.* **322** 92–96
[92] Desgranges C and Delhommelle J 2012 *J. Chem. Phys.* **136** 184107
[93] Desgranges C and Delhommelle J 2012 *J. Chem. Phys.* **136** 184108
[94] Ngale K N, Desgranges C and Delhommelle J 2012 *Mol. Sim.* **38** 653–658
[95] Hicks J M, Desgranges C and Delhommelle J 2012 *J. Phys. Chem. C* **116** 22938–22946
[96] Rathore N, IV T A K and de Pablo J J 2003 *J. Chem. Phys.* **118** 4285–4290
[97] Rathore N, IV T A K and de Pablo J J 2006 *Biophys. J.* **90** 1767–1773
[98] Frenkel D and Smit B 2002 *Understanding Molecular Simulation: From Algorithms to Applications* (San Diego, CA: Academic Press)
[99] Lima A R, de Oliveira P M C and Penna T J P 2000 *J. Stat. Phys.* **99** 691–705
[100] Paluch A S, Shen V K and Errington J R 2008 *Ind. Eng. Chem. Res.* **47** 4533–4541
[101] Shen V K and Errington J R 2005 *J. Chem. Phys.* **122** 064508
[102] Errington J R and Shen V K 2005 *J. Chem. Phys.* **123** 164103
[103] Singh J K and Errington J R 2006 *J. Phys. Chem. B* **110** 1369–1376
[104] Rosch T W and Errington J R 2007 *J. Phys. Chem. B* **111** 12591–12598
[105] Singh J K and Kwak S K 2007 *J. Chem. Phys.* **126** 024702
[106] Rosch T W and Errington J R 2008 *J. Chem. Phys.* **129** 164907
[107] Kumar A N and Singh J K 2008 *Mol. Phys.* **106** 2277–2288
[108] Singh S K, Sinha A, Deo G and Singh J K 2009 *J. Phys. Chem. C* **113** 7170–7180
[109] Jana S, Singh J K and Kwak S K 2009 *J. Chem. Phys.* **130** 214707
[110] Singh S K, Sinha A and Singh J K 2010 *J. Phys. Chem. B* **113** 4283–4292
[111] Chen H and Sholl D S 2006 *Langmuir* **22** 709–716

[112] Chen H and Sholl D S 2007 *Langmuir* **23** 6431–6437
[113] Errington J R 2004 *Langmuir* **20** 3798–3804
[114] Grzelak E M and Errington J R 2008 *J. Chem. Phys.* **128** 014710
[115] Grzelak E M, Shen V K and Errington J R 2010 *Langmuir* **26** 8274–8281
[116] Grzelak E M and Errington J R 2010 *J. Chem. Phys.* **132** 224702
[117] Grzelak E M and Errington J R 2010 *Langmuir* **26** 13297–13304
[118] Kumar V, Sridhar S and Errington J R 2011 *J. Chem. Phys.* **135** 184702
[119] Wilding N B 2001 *Am. J. Phys.* **69** 1147–1155
[120] Paluch A S, Jayaraman S, Shah J K and Maginn E J 2010 *J. Chem. Phys.* **133** 124504
[121] Potoff J J and Panagiotopoulos A Z 1998 *J. Chem. Phys.* **109** 10914–10920
[122] Martin M G and Siepmann J I 1998 *J. Phys. Chem. B* **102** 2569–2577
[123] Siepmann J I, Karaborni S and Smit B 1993 *Nature* **365** 330–332
[124] Wood W W and Parker F R 1957 *J. Chem. Phys.* **27** 720–733
[125] Rowlinson J S and Widom B 1989 *Molecular Theory of Capillarity* (New York, NY: Oxford University Press)
[126] Rowlinson J S and Swinton F L 1982 *Liquids and Liquid Mixtures* (London: Butterworth)
[127] Schultz B J, Binder K, Müller M and Landau D P 2003 *Phys. Rev. E* **67** 067102
[128] Tröster A and Dellago C 2005 *Phys. Rev. E* **71** 066705
[129] Escobedo F A and Abreu C R A 2006 *J. Chem. Phys.* **124** 104110
[130] Cunha-Netto A G, Caparica A A, Tsai S H, Dickman R and Landau D P 2008 *Phys. Rev. E* **78** 055701
[131] Tsai S H, Wang F and Landau D P 2008 *Braz. J. Phys.* **38** 6–11
[132] Shell M S, Debenedetti P G and Panagiotopoulos A Z 2004 *J. Phys. Chem. B* **108** 19748–19755
[133] NIST Chemistry WebBook, NIST Standard Reference Database Number 69, Eds. P.J. Linstrom and W.G. Mallard, National Institute of Standards and Technology, Gaithersburg MD, 20899 (http://webbook.nist.gov)
[134] Agrawal R and Kofke D A 1995 *Mol. Phys.* **85** 43–59
[135] Chen B and Siepmann J I 2001 *J. Phys. Chem. B* **105** 9840–9848
[136] Heng V R, Nayhouse M, Crose M, Tran A and Orkoulas G 2012 *J. Chem. Phys.* **137** 141101
[137] Chen B and Siepmann J I *in preparation*
[138] MacDowell L G, Shen V K and Errington J R 2006 *J. Chem. Phys.* **125** 034705
[139] Godawat R, Jamadagni S N, Errington J R and Garde S 2008 *Ind. Eng. Chem. Res.* **47** 3582–3590
[140] Binder K, Block B J, Virnau P and Troöster A 2012 *Am. J. Phys.* **80** 1099–1109

A Conformational Search Method for Protein Systems Using Genetic Crossover and Metropolis Criterion

Yoshitake Sakae[1,2], **Tomoyuki Hiroyasu**[3], **Mitsunori Miki**[4], **Katsuya Ishii**[5], **and Yuko Okamoto**[1,5,6,7]

[1] Department of Physics, Graduate School of Science, Nagoya University, Nagoya, Aichi 464-8602, Japan
[2] Computational Materials Science Initiative, Institute for Molecular Science, Okazaki, Aichi, 444-8585, Japan
[3] Department of Biomedical Information, Doshisha University, Kyotanabe, Kyoto 610-0394, Japan
[4] Department of Intelligent Information Engineering and Sciences, Doshisha University, Kyotanabe, Kyoto 610-0394, Japan
[5] Information Technology Center, Nagoya University, Nagoya, Aichi 464-8601, Japan
[6] Structural Biology Research Center, Graduate School of Science, Nagoya University, Nagoya, Aichi 464-8602, Japan
[7] Center for Computational Science, Graduate School of Engineering, Nagoya University, Nagoya, Aichi 464-8603, Japan

E-mail: `okamoto@phys.nagoya-u.ac.jp`

Abstract. Many proteins carry out their biological functions by forming the characteristic tertiary structures. Therefore, the search of the stable states of proteins by molecular simulations is important to understand their functions and stabilities. However, getting the stable state by conformational search is difficult, because the energy landscape of the system is characterized by many local minima separated by high energy barriers. In order to overcome this difficulty, various sampling and optimization methods for conformations of proteins have been proposed. In this study, we propose a new conformational search method for proteins by using genetic crossover and Metropolis criterion. We applied this method to an α-helical protein. The conformations obtained from the simulations are in good agreement with the experimental results.

1. Introduction

To understand functions and stabilities of biomolecules such as DNA and proteins, effective conformational search and accurate estimation of canonical distributions are important. However, because biomolecules have a lot of local minimum-energy states separated by high energy barriers, conventional molecular dynamics (MD) and Monte Carlo (MC) simulations tend to get trapped in states of local minima. In order to overcome this difficulty, various sampling and optimization methods for conformations of biomolecules have been proposed such as generalized-ensemble algorithms (for reviews, see, e.g., [1, 2]). Moreover, we have proposed a conformational search method called the parallel simulated annealing using genetic crossover

(PSA/GAc) [3, 4, 5, 6], which is a hybrid algorithm combining both the simulated annealing (SA) [7] and the genetic algorithm (GA) [8, 9]. In this method, a simulated annealing simulation is combined with genetic crossover, which is one of the operations of genetic algorithm.

In this paper, we propose a new conformational search method that combines the genetic crossover and Metropolis criterion. The operation of the genetic crossover is combined with the conventional MC or MD simulations. In order to examine the effectiveness of our method, we applied our method to a small protein, which is the 10-55 helical fragment B of protein A from *Staphylococcus aureus* (in this paper, just referred to as protein A) [10]. The results of this simulation are found to be in good agreement with the experimental results.

2. Methods

In the present simulation approach, we first prepare M initial conformations of the system in study, where M is the total number of "individuals" in genetic algorithm and is usually taken to be an even integer. We then alternately perform the following two steps:

1. For the M individuals, regular canonical MC or MD simulations at a fixed temperature T are carried out simulataneously and independently for a certain MC or MD steps.
2. $M/2$ pairs of conformations are selected from "parental" group randomly, and the crossover and selection operations are performed. Here, the parental group means the latest conformations obtained in Step 1.

While Step 1 is usually based on local updates of conformations, global updates of conformations are introduced in Step 2 by genetic crossover. The latter greatly enhances the conformational sampling.

In the following, we give the details of Step 2 above. We can employ various kinds of genetic crossover operations such as one-point crossover [3, 4, 5], two-point crossover [6], etc. as in our previous simulation methods. Here, we just present a case of the two-point genetic crossover. The crossover operation in this method exchanges a part of corresponding dihedral angles between two conformations of the protein. In the two-point crossover operation on a parental pair, the following procedure is carried out:

(i) Consecutive amino acids of length n residues in the amino-acid sequence of the conformation are selected randomly for each pair of selected conformations.

(ii) Dihedral angles (in only backbone or all dihedral angles) in the selected n amino acids are exchanged between the selected pair of conformations.

Note that the length n of consecutive amino-acid residues can, in general, be different for each pair of selected conformations.

We need to deal with the produced "child" conformations with care. Because the produced conformations often have unnatural structures by the crossover operation, they have high potential energy and are unstable. Therefore, the relaxation process is introduced before the selection operation. In this paper, we use equlibration simulations with restraints as a relaxation process. For instance, short simulations at the same temperature T with restraints on the backbone dihedral angles of only the n amino acids are performed so that the corresponding backbone structures of the n amino acids will approach the exchanged backbone conformation. The initial conformations for these equilibration simulations are the ones before the exchanges. Namely, by these equilibration simulations, the corresponding backbone conformations of the n amino acids gradually transform from the ones before the exchanges to the ones after the exchanges. We then perform short equilibration simulations without the restraints. We select the last conformations in the equilibratoin simulations as "child" conformations. Note that from a pair of parental conformations we get two child conformations. In the present method, we consider selection between parent and child for the parent-child pair with the same conformations in the remaining amino acids (other than n consecutive ones).

In the final stage in Step 2, the selection operation is performed. We select a superior "chromosome" (conformation) from the parent-child pair. For this selection operation, we can also employ various criteria. In this study, we employ Metropolis criterion, which selects the new child conformation from the parent with the following probability:

$$w(\mathrm{p} \to \mathrm{c}) = \min\left(1, \exp\{-\beta[E_\mathrm{c} - E_\mathrm{p}]\}\right). \tag{1}$$

Here, E_p and E_c stand for the potential energy of the parental conformation and the child conformation, respectively, of the parent-child pair. β is the inverse temperature, which is defined by $\beta = 1/k_\mathrm{B}T$ (k_B is the Boltzmann constant).

We remark that if we use different temperatures, we can also introduce the replica-exchange method [11, 12] together with the above method in order to further enhance conformational sampling [13].

3. Results and Discussion

We applied the present method to protein A. Although the whole protein A has 60 amino acids, we used the truncated 46 amino-acid sequence from Gln10 to Ala55. For this simulation, we used the AMBER12 program package and incorporated the two-point genetic crossover procedure. The unit time step was set to 2.0 fs and the bonds involving hydrogen atoms were constrained by the SHAKE algorithm [14]. Each simulation for sampling was carried out for 90.0 nsec (which consisted of 45,000,000 MD steps) with 32 individuals ($M = 32$) and performed the crossover operations 90 times during the simulation. The temperature during the simulations was kept at 300 K by using Langevin dynamics. The nonbonded cutoff of 20 Å was used. As for solvent effects, we used the GB/SA model [15] included in the AMBER12 program package ($igb = 5$). In the crossover operations, we set the length n of consecutive amino-acid residues to be an even integer ranging from 10 to 20. This number was chosen randomly for each pair of parental conformations. As for the equilibration simulations just after each genetic crossover operation, the first simulations with the harmonic restraints on the backbone diheral angles of n amino-acid residues (the force constants were 600 kcal/mol·Å^2) lasted for 20 psec, and the following simulations without restraints also lasted for 20 psec.

We obtained a similar conformation to the experimental native structure, and its root-mean-square distance (RMSD) (for only the backbone atoms) from the native structure was 1.7 Å (see Fig. 1). One of the characteristic analyses of the genetic algorithm is the analysis of data as functions of the generation number. In this paper, the generation number stands for the number of crossover operations. In Fig. 2, we show four kinds of values, the fraction of helix conformations, the average of the potential energy, the average of the RMSD, and the average of the radius of gyration by changing the generation. In Fig. 2(a), we see that the fraction of helix conformations increases as the generation proceeds as a whole. The PDB structure of protein A has three helix regions in the amino-acid sequence of 2–9, 17–29, and 34–46. These helix regions roughly correspond to the high probability regions obtained from the present simulation, and the C-terminal region (Helix III) has the highest helix fraction among the three regions in Fig. 2(a). The stabilities for the three helix structures in protein A have been examined and debated by many researchers [16, 17, 18, 19, 20]. From our simulation results, we consider that Helix III is the most stable, and the central helix is unstable in comparison with the other two helix structures. In Fig. 2(b)–(d), we see that the potential energy, RMSD, and radius of gyration quickly decrease at the early numbers of generations. After that, although these values are roughly convergent, there are some fluctuations. For example, RMSD value has the standard deviation of 1.7 Å around a mean value of about 8.0 Å.

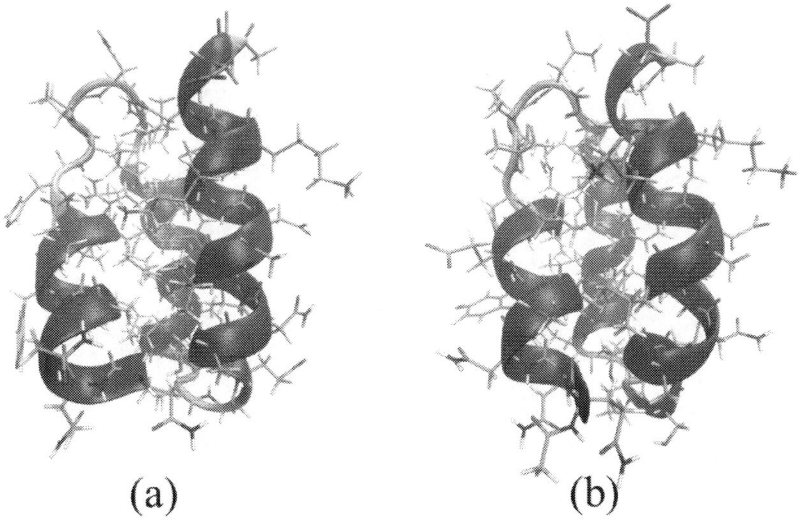

Figure 1. Structures of protein A. (a) PDB structure (PDB ID: 1BDD). (b) A conformation obtained from the present simulation, which has the lowest RMSD value from the PDB structure (RMSD = 1.7 Å).

4. Conclusions

In this work, we proposed a new conformational search method for protein systems. This method combines conventional canonical MC or MD simulations, which are based on local updates of conformations, and genetic crossover, which is based on global updates of conformations. The latter greatly enhances the conformational sampling. In order to examine the efficiency of this method, we applied it to protein A. We found conformations very close to the native structure.

The present method is particularly suitable for highly parallel computers. In the future, we are going to apply this method to larger protein systems.

Acknowledgments

The computations were performed on the computers at the Research Center for Computational Science, Institute for Molecular Science, Information Technology Center, Nagoya University, and Center for Computational Sciences, University of Tsukuba. This work was supported, in part, by the Grants-in-Aid, for Scientific Research on Innovative Areas ("Dynamical Ordering & Integrated Functions"), for the Computional Materials Science Initiative, and for High Performance Computing Infrastructure, from the Ministry of Education, Culture, Sports, Science and Technology (MEXT), Japan.

References

[1] Hansmann U and Okamoto Y 1999 *Curr. Opin. Struct. Biol.* **9** 177–183

[2] Mitsutake A, Sugita Y and Okamoto Y 2001 *Biopolymers* **60** 96–123

[3] Hiroyasu T, Miki M and Ogura M 2000 *Proceedings of the 44th Institute of Systems* 113–114

[4] Hiroyasu T, Miki M, Ogura M and Okamoto Y 2002 *J. IPS Japan* **43** 70–79

[5] Hiroyasu T, Miki M, Ogura M, Aoi K, Yoshida T and Okamoto Y 2003 *Proceedings of the 7th World Multiconference on Systemics, Cybernetics and Informatics (SCI 2003)* 117–122

[6] Sakae Y, Hiroyasu T, Miki M and Okamoto Y 2011 *J. Comput. Chem.* **32** 1353–1360

[7] Kirkpatrick S, Gelatt Jr C D and Vecchi M P 1983 *Science* **220** 671–680

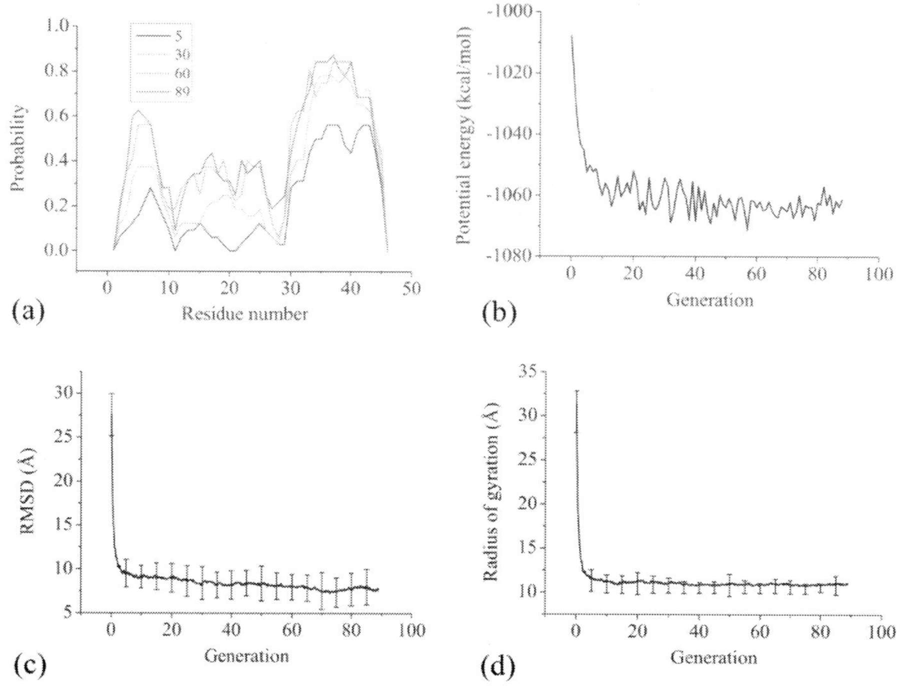

Figure 2. (a) The fraction of helix conformations as a function of residue number at generation numbers 5 (blue), 30 (light blue), 60 (green), and 89 (red), (b) the average of the potential energy, (c) the average of the RMSD, and (d) the average of the radius of gyration, obtained from the present GC simulation results. The helix conformations were examined by using the DSSP program [21]. The error bars in (c) and (d) are standard deviations.

[8] Holland J H 1975 *Adaptation in Natural and Artificial Systems* (Ann Arbor: The University of Michigan Press)

[9] Goldberg D E 1989 *Genetic Algorithms in Search, Optimization, and Machine Learning* (Reading: Addison-Wesley)

[10] Gouda H, Torigoe H, Saito A, Sato M, Arata Y and Shimada I 1992 *Biochemistry* **31** 9665–9672

[11] Hukushima K and Nemoto K 1996 *J. Phys. Soc. Jpn.* **65** 1604–1608

[12] Sugita Y and Okamoto Y 1999 *Chem. Phys. Lett.* **314** 141–151

[13] Sakae Y, Hiroyasu T, Miki M, Ishii K and Okamoto Y manuscript in preparation

[14] Ryckaert J P, Ciccotti G and Berendsen H J C 1977 *J. Comput. Phys.* **23** 327–341

[15] Onufriev A, Bashford D and Case D A 2004 *Proteins* **55** 383–394

[16] Zhou Y and Karplus M 1999 *Nature* **401** 400–403

[17] Alonso D O V and Daggett V 2000 *Proc. Natl. Acad. Sci. U.S.A.* **97** 133–138

[18] Garcia A E and Onuchic J N 2003 *Proc. Natl. Acad. Sci. U.S.A.* **100** 13898–13903

[19] Sato S, Religa T L, Daggett V and Fersht A R 2004 *Proc. Natl. Acad. Sci. U.S.A.* **101** 6952–6956

[20] Shao Q and Gao Y Q 2011 *J. Chem. Phys.* **135** 135102

[21] Kabsch W and Sander C 1983 *Biopolymers* **22** 2577–2637

Journal of Physics: Conference Series **487** (2014) 012004 doi:10.1088/1742-6596/487/1/012004

On ripples and rafts: Curvature induced nanoscale structures in lipid membranes

Friederike Schmid[1], **Stefan Dolezel** [1], **Olaf Lenz**[2], **Sebastian Meinhardt**[1]

[1] Institut für Physik, Johannes Gutenberg-Universität Mainz, DE
[2] ICP, Universität Stuttgart, DE

E-mail: `friederike.schmid@uni-mainz.de`

Abstract.
We develop an elastic theory that predicts the spontaneous formation of nanoscale structures in lipid bilayers which locally phase separate between two phases with different spontaneous monolayer curvature. The theory rationalizes in a unified manner the observation of a variety of nanoscale structures in lipid membranes: Rippled states in one-component membranes, lipid rafts in multicomponent membranes. Furthermore, we report on recent observations of rippled states and rafts in simulations of a simple coarse-grained model for lipid bilayers, which are compatible with experimental observations and with our elastic model.

1. Introduction

Biomembranes are essential components of all living matter, and lipids are essential components of biomembranes. According to the venerable "fluid mosaic model" due to Singer and Nicolson [1], the basic structure of biomembranes is provided by a self-assembled lipid bilayer, which serves as a matrix to carry proteins which then perform various biological functions. In the last decades, it has become clear that biomembranes are not laterally homogeneous, but internally structured ("dynamically structured mosaic model" [2]), and this structure is believed to be important for many membrane-associated biological processes. In particular, the lipid raft hypothesis asserts that biomembranes are often filled with nanoscopic domains characterized by a different lipid composition – typically a higher cholesterol content – and higher local order [3–8]. There is ample indirect evidence for the existence of such domains, e.g., from diffusion experiments [9, 10], or from superresolution fluorescence images showing nanoscopic clusters of raft-associated proteins in membranes [10–12]. Nevertheless, the lipid raft hypothesis is still discussed controversially [8]. For example, it is argued that the observed clustering of proteins could be driven by proteins only.

The lipid raft concept is also intriguing from a theoretical point of view. Multicomponent lipid bilayers are known to phase separate in certain parameter regions into a "liquid disordered" (ld) and a "liquid ordered" (lo) phase [13–15], and this has also been observed in plasma membranes [16]. The hypothetical "raft state", however, is not phase-separated, but rather believed to be a globally homogeneous state filled with nanodomains of sizes between 10 and 100 nanometers. To explain this, it was proposed that rafts might be nonequilibrium structures [17] which are prevented from merging by kinetic barriers [18]. Others have suggested that finite-

Journal of Physics: Conference Series **487** (2014) 012004 doi:10.1088/1742-6596/487/1/012004

size equilibrium rafts might be stabilized by the cytoplasm [19], by special line-active lipids or proteins [20–23], or by elastic interactions arising from local composition differences in apposing monolayer leaflets [24] (the last mechanism requires that the membrane is under tension). Alternatively, it was argued that "rafts" might simply be a signature of critical fluctuations in the vicinity of critical points [25, 26].

Rafts are not the only nanoscale structures in lipid membranes. Many one-component lipid membranes exhibit a periodically modulated "ripple" phase ($P_{\beta'}$) in the transition region between the high-temperature fluid phase (L_α) and the low-temperature gel phase. The characteristic wave length of the ripple phase (15 nm [27]) is of the same order than that typically attributed to lipid rafts (10-100 nm). Experimentally [28, 29] and in computer simulations [30–35], modulated phases are observed in lipid bilayers that exhibit a tilted gel state, and they are not observed in lipid bilayers with an untilted gel state L_β [28–30, 36]. For example, we have introduced a simple generic lipid model [37–39] which reproduces the L_α phase and the $L_{\beta'}$ phase as well as the ripple phase [32], it the size of the head groups is chosen sufficiently large. However, if we reduce the head size such that the tilt in the gel phase vanishes, the ripple phase disappears as well [36].

These findings suggest that the ripple phase emerges as a response to internal stresses in the membrane, which are generated by a mismatch of head and tail group. At low temperatures, the lipids adjust to the mismatch by uniformly tilting in one direction. This partially relieves the stress, but the remaining stress profile is still substantial [40]. More efficient stress relaxation can be achieved if the monolayers curve around, such that lipids can splay. As domains with uniform monolayer curvature obviously cannot extend to infinity, this might explain the formation of gel stripes with finite width separated by small lines of fluid phase [32].

A similar mechanism could stabilize lipid rafts. Indeed, we have recently carried out simulations of a multicomponent version of the generic lipid membrane model mentioned above [39], and found a structure which bears all the characteristics of rafts. The membranes assume an in-plane structure which is akin to a two-dimensional microemulsion with nanoscale domains of "liquid ordered" phase embedded in a sea of "liquid disordered phase" [41]. We have also developed a simple elastic theory which rationalizes both rippled states and rafts in a unified manner.

This work is reviewed in the present conference proceedings paper. We will first present the theory sketched in Ref. [41] in some more detail. Then we will recapitulate some main results of the simulations, both of the ripple phase and of lipid rafts, that support the theory. We conclude with a brief summary.

2. Theory

Our starting point is an elastic theory for membranes that describes the lipid bilayer as a system of two coupled elastic monolayers [42–46]. The phase transition inside the membrane (liquid / gel or liquid ordered / liquid disordered, respectively), is described in a Ginzburg-Landau fashion by an internal order parameter Φ, which is taken to be equal on both monolayer leaflets. Hence the bilayer is perfectly symmetric also in composition. This is clearly an oversimplification. In the raft case, it can be motivated by the experimental observation that *globally* phase separated liquid ordered domains are strongly correlated across bilayers and tend to oppose each other [47]. In the ripple case, the liquid domains typically are typically arranged in a staggered order. This has consequences on the shape of the total membrane (it assumes a zigzag structure), which will be investigated in future work.

For perfectly symmetric membranes, the bilayer remains on average flat and the total free energy can be described by a combination of two terms, $\mathcal{F} = \mathcal{F}_{\text{el}} + \mathcal{F}_\Phi$, where $\mathcal{F}_{\text{el}}$ describes the elastic energy of the coupled monolayer system, and $\mathcal{F}_\Phi$ the Ginzburg-Landau energy associated with the order parameter Φ. In Monge representation[48], the elastic energy can be written as

[46]

$$\mathcal{F}_{\rm el} = \int {\rm d}^2r \Big\{ \frac{k_c}{2}(\Delta u)^2 + \frac{k_A}{2t_0^2}u^2 + 2k_c(c_0 + \frac{\zeta}{t_0}u)\Delta u + k_G \det(\partial_{ij}u) \Big\}, \qquad (1)$$

where $u(\mathbf{r})$ denotes the local deviation from the mean monolayer thickness t_0, and the other parameters characterize the elastic properties of the membrane: k_c the (bilayer) bending stiffness, k_A the (bilayer) compression modulus, c_0 the spontaneous monolayer curvature, ζ an associated curvature-related parameter [44, 45], and k_G the Gaussian curvature of monolayers. At fixed lipid number and membrane tension, Eq. (1) holds both for tensionless membranes and membranes under tension [49]. (However, elastic constants may depend on the tension [49, 50]). At fixed membrane area and variable lipid number, an additional term $\mu_l(\nabla u)^2$ comes in for membranes under tension [51], which is associated with the chemical potential of the lipids, and can be incorporated in the parameter ζ after partial integration. We have used Eq. (1) several times in the past to fit deformation profiles in the vicinity of inclusions, with good results down to molecular length scales [46, 52].

The Ginzburg-Landau associated with the order parameter Φ is given by

$$\mathcal{F}_{\Phi} = \int {\rm d}^2r \Big\{ \frac{g}{2}(\nabla\Phi)^2 + f(\Phi) \Big\}, \qquad (2)$$

where $f(\Phi)$ is a local free energy functional with two minima that account for the local tendency of phase separation. The elastic degrees of freedom $u(\mathbf{r})$ are coupled with the order parameter Φ through the fact that elastic parameters may vary spatially and depend on $\Phi(\mathbf{r})$ [53]. For example, only recently [54] it was discussed how the monolayer curvature c_0 might depend on the membrane composition for raft-forming lipids.

Here, we will assume for simplicity that all elastic parameter are constant except the monolayer curvature, c_0, which is taken to depend on Φ in a linear fashion. In this case, the natural length unit in the system is the in-plane correlation length $\xi = (k_c t_0^2/k_A)^{1/4}$, which is the length scale that one can construct from combining the bending rigidity k_c and the reduced compression modulus k_A/t_0^2 (a similar expression is found in membrane stacks [55, 56]). Micropipette experiments on a variety of lipid membranes [57] have shown that k_c is roughly proportonal to $k_A t_0^2$, hence ξ should be of the order of the membrane thickness, t_0. Inserting actual numbers for the fluid phase of one-component bilayers made of dipalmitoyl phosphatidylcholine (DPPC, one of the most common lipids in natural biomembranes) from experiments, all-atom simulations, or coarse-grained simulations [46], one consistently obtains values around $\xi \sim (0.9 - 1.4)nm$.

Rescaling all lengths in units of ξ, the elastic energy, Eq. (1), can be rewritten as

$$\mathcal{F}_{\rm el} = k_c \int {\rm d}^2\hat{r} \Big\{ \frac{l}{2}(\hat{\Delta}\hat{u})^2 + \frac{1}{2}\hat{u}^2 + (2\Phi\hat{c} + b\hat{u})\hat{\Delta}\hat{u} \Big\}, \qquad (3)$$

with the dimensionless characteristic membrane parameter $b = 2\zeta\xi^2/t_0$ ($b \sim 0.7$ in DPPC[41]) and $\hat{c} = \xi c_0(\Phi)/\Phi =$ const. The contribution of the Gaussian rigidity, k_G, has been dropped, because it gives a constant at fixed topology according to the Gauss-Bonnet theorem. The Ginzburg-Landau energy is rescaled accordingly,

$$\mathcal{F}_{\Phi} = \int {\rm d}^2\hat{r} \Big\{ \frac{g}{2}(\hat{\nabla}\Phi)^2 + \hat{f}(\Phi) \Big\}, \qquad (4)$$

with $\hat{f}(\Phi) = \xi^2 f(\Phi)$.

2.1. Weak segregation limit and Landau-Brazovskii approximation

We first discuss the weak segregation case, where local deviations from a homogeneous reference state are small and Φ remains close to zero. We minimize the elastic free energy, $\mathcal{F}_{\rm el}$, with respect to the bilayer distortion $\hat{u}(\hat{r})$. This gives the Euler Lagrange equations

$$\hat{\Delta}^4\hat{u} + 2b\hat{\Delta}\hat{u} + \hat{u} + 2\hat{c}\hat{\Delta}\Phi = 0 \tag{5}$$

The solution is most conveniently calculated in Fourier space,

$$\hat{u}_{\mathbf{q}} = \frac{2\hat{c}q^2}{q^4 - 2bq^2 + 1}\,\Phi_{\mathbf{q}}, \tag{6}$$

where the $\mathbf{q}$ are in-plane wavevectors. Inserting this expression in Eq. (4) and dropping boundary terms, one obtains

$$\mathcal{F} = \frac{1}{2}\sum_{\mathbf{q}} |\Phi_{\mathbf{q}}|^2 q^2 g_{\rm eff}(q^2) + \int \mathrm{d}^2\hat{r}\, \hat{f}(\Phi) \tag{7}$$

with

$$g_{\rm eff}(q^2) = g - 4k_c\hat{c}^2\frac{q^2}{q^4 - 2bq^2 + 1}. \tag{8}$$

Inspection of Eq. (7) shows that a homogeneous phase ($\Phi \equiv$ const., i.e., $\Phi_{\mathbf{q}} = 0$ for $\mathbf{q} \neq 0$) becomes unstable with respect to modulations with wavevector $q^* = 1$ (i.e., wavelength ξ) if the stiffness parameter g becomes smaller than $g \leq g^* = 2k_c\hat{c}^2/(1-b)$. Close to the transition, modulations with $q \approx q^*$ dominate, and we can expand $q^2 g_{\rm eff}(q^2)$ about $q^2 \approx q^{*2}$ up to second order in $(q^2 - q^{*2})$. Furthermore, we expand the local free energy density $\hat{f}(\Phi)$ in powers of Φ up to fourth order,

$$\hat{f}(\Phi) = \frac{a}{2}\Phi^2 - \frac{\gamma}{6}\Phi^3 + \frac{\lambda}{24}\Phi^4. \tag{9}$$

Putting everything together and transforming the stiffness contribution $\sum_{\mathbf{q}} |\Phi_{\mathbf{q}}|^2 q^2 g_{\rm eff}(q^2)$ back to real space, our final free energy functional for Φ can be written in the form

$$\mathcal{F} = \int \mathrm{d}^2\hat{r}\Big\{\frac{\Gamma}{8q_0^2}(\hat{\Delta} + q_0^2)^2\Phi^2 + \frac{\tau}{2}\Phi^2 - \frac{\gamma}{6}\Phi^3 + \frac{\lambda}{24}\Phi^4\Big\} \tag{10}$$

with $q_0^2 = (q^*)^2\,(1 + (1-b)(1 - g/g^*))$, $\Gamma/4q_0^2 = g^*/2(1-b)$, and $\tau = a - (g - g^*)(1 + q_0^2)/2$. Eq. 10) is the free energy of the well-known Landau-Brazovskii model [58], which provides a general framework for the description of phase transitions driven by a short wavelength instability between a disordered phase and one or several ordered phases. In mean field approximation, the Brazovskii model predicts a continuous transition point between a homogeneous phase and several ordered phases (stripes, hexagonal). Fluctuations are known to destroy the order close to the multicritical point and stabilize a locally structured disordered phase *via* the so-called Brazovskii mechanism [58]. The order-disorder transition is shifted and becomes first order. Hence this model accounts for both disordered states with local structure and a characteristic wave length, such as raft states, and ordered modulated states such as ripple phases.

2.2. Strong segregation limit and raft domains

We can also approach the problem from the opposite side and consider membranes where the local tendency for phase separation is very strong, i.e., domains separated by narrow interfaces form throughout the system . Kuzmin et al. have pointed out in 2005 that curvature effects reduce the line tension between such domains [18]. Here, we will show that they may even reduce the line tension to zero, such that globally phase separated states are no longer stable, and stabilize rafts of finite size.

We assume that the interfaces between the phase separated domains are sufficiently narrow that they can be treated as lines where the order parameter jumps from one value to another, with a bare line tension λ_0. The spontaneous curvature c_0 in the two phases differs and the difference is denoted δc_0. Within each domain, the elastic parameters are homogeneous, and the solution for $\hat{u}(\hat{r})$ that minimizes the elastic free energy $\mathcal{F}_{\text{el}}$ must fulfill the homogeneous Euler-Lagrange equation,

$$\hat{\Delta}^4 \hat{u} + 2b\hat{\Delta}\hat{u} + \hat{u} = 0. \tag{11}$$

The solution depends on the boundary condition. Since the equation is fourth order, four sets of boundary conditions have to be given, e.g., equations for $\hat{u}(\mathbf{r})$ and the first three derivatives of $\hat{u}(\mathbf{r})$ along the boundary normals $\mathbf{n}$. These boundary conditions must be chosen such that they minimize the total elastic energy. In particular, it is clear from Eq. (3) that $\hat{u}(\mathbf{r})$ and its normal derivative, $\mathbf{n}\cdot\hat{\nabla}\hat{u}$, must be continous at the boundaries, otherwise the bending contribution $\int \frac{1}{2}(\hat{\Delta}\hat{u})^2$ would diverge.

To derive the remaining boundary conditions, we carry out a variation of $\mathcal{F}_{\text{el}}$ with free boundary conditions for all boundaries $\mathcal{C}_i$ of domains i with boundary normals $\mathbf{n}_i$ (pointing outwards) and spontaneous curvatures $c_0^{(i)}$. This leads to the variational expression $\delta\mathcal{F}_{\text{el}} = \delta\mathcal{F}_{\text{el},b} + \delta\mathcal{F}_{\text{el},s}$ with

$$\delta\mathcal{F}_{\text{el},b} \propto \int \mathrm{d}^2\hat{r} \Big[\hat{\Delta}^4\hat{u} + 2b\hat{\Delta}\hat{u} + \hat{u}\Big]\delta\hat{u} \tag{12}$$

and

$$\delta\mathcal{F}_{\text{el},s} \propto \int_{\mathcal{C}_i} \mathrm{d}l \,\Big\{\delta(\mathbf{n}_i\cdot\hat{\nabla}\hat{u})[\hat{\Delta}\hat{u} + b\hat{u} + 2\xi c_0^{(i)}] - \delta\hat{u}[(\mathbf{n}_i\cdot\hat{\nabla})\hat{\Delta}\hat{u} + b(\mathbf{n}_i\cdot\hat{\nabla})\hat{u}]\Big\}. \tag{13}$$

The functional $\delta\mathcal{F}_{\text{el}}$ has to vanish for all possible variations $\delta\hat{u}(\mathbf{r})$ in the bulk and for all variations $\delta\hat{u}$ and $\delta(\mathbf{n}\cdot\hat{\nabla}\hat{u})$ at the boundaries. The bulk term $\delta\mathcal{F}_{\text{el},b}$, gives the Euler-Lagrange equations. The surface term, $\delta\mathcal{F}_{\text{el},s}$, is used to derive the appropriate boundary conditions. In our case, the membrane is completely filled with domains, hence every boundary integral element dl appears twice with normals pointing in opposite directions. Since, furthermore, $\hat{u}$ and $\mathbf{n}\cdot\hat{\nabla}\hat{u}$ must be continuous at the boundaries, a number of terms drop out and we are left with the boundary conditions that $\hat{\nabla}\hat{\Delta}\hat{u}$ and $(\hat{\Delta}\hat{u} + 2\xi c_0)$ must be continuous.

Inserting these boundary conditions and the Euler-Lagrange equations in $\mathcal{F}_{\text{el}}$ (Eq. (3), we obtain the following simple expression for the elastic energy,

$$\mathcal{F}_{\text{el}} = k_c\,\delta c_0 \int \mathrm{d}l\; \mathbf{n}\cdot\hat{\nabla}\hat{u}, \tag{14}$$

where the line integral $\int \mathrm{d}l$ runs once over all domain boundaries, $\mathbf{n}$ is the normal vector, δc_0 is the curvature mismatch, i.e., the monolayer curvature difference in the inner and outer phase, and u is the solution ($u = \xi\hat{u}$) of the Euler-Lagrange equation with the appropriate boundary conditions.

The rest of the calculation is purely technical. In a one-dimensional stripe geometry, the general solution of the Euler Lagrange equations is given by an exponential

$$\hat{u}(\hat{z})) = e^{\pm i\alpha\hat{z}} \qquad \text{with} \quad \alpha = \sqrt{\frac{1}{2}}(\sqrt{1+b} + i\sqrt{1-b}). \tag{15}$$

Inserting the boundary conditions, one obtains the following solution for a single stripe domain of width D:

$$u(z) = -\frac{2\xi^2\,\delta c_0}{\sqrt{1-b^2}} \begin{cases} \Re[ie^{i\alpha D/2}\cos(\alpha z)] & : 0 < z < D/2 \\ \Re[\sin(-\alpha D/2)\,e^{i\alpha z}] & : z > D/2 \end{cases} \tag{16}$$

with $u(-z) = u(z)$. The corresponding elastic free energy is $\mathcal{F}_{el} = 2L\lambda_{el}$ (L is the length of the stripe) with

$$\lambda_{el}(D) = -\frac{\xi k_c \delta c_0^2}{\sqrt{1-b^2}} \Re[\alpha(1-e^{i\alpha D})]. \tag{17}$$

We note that this expression is always negative. Relaxing the elastic energy at the boundaries goes along with a gain in free energy. The total energy of a stripe, which includes the bare tension λ_0, is given by $\mathcal{F}_{el} = 2L(\lambda_0 + \lambda_{el})$, and it is reduced by the contribution of the elastic energy. In particular, in the limit $D \to \infty$, one obtains the line tension

$$\lambda = \lambda_0 + \lambda_{el}(\infty) \qquad \text{with} \quad \lambda_{el}(\infty) = -\frac{\xi k_c \delta c_0^2}{\sqrt{2(1-b)}}. \tag{18}$$

Hence the contribution due to the monolayer is found to reduce the line tension, in agreement with Kuzmin et al. [18]. At high curvature mismatch, the line tension may become negative. Even more interestingly, it exhibits an oscillatory behavior as a function of stripe width (Eq. 17), indicating that stripes of finite width are favored. The optimal width is of order 4ξ, i.e., several nanometers, which is just the width of ripples and the width commonly attributed to rafts.

A similar exercise can be carried out for domains with spherical geometries. In that case, we look for radially symmetric solutions of the Euler Lagrange equations (11), which can be written in the form

$$u(r) = \frac{\delta c_0}{\sqrt{1-b^2}} \frac{\pi D \xi}{2} \begin{cases} \Re[\xi\alpha\, J_0(\alpha r)\, H_1^{(1)}(\alpha D/2)] & : r < R \\ \Re[\xi\alpha\, J_1(\alpha D/2)\, H_0^{(1)}(\alpha r)] & : r > R \end{cases}, \tag{19}$$

where J_n and $H_n^{(1)}$ are the Bessel and Hankel functions of the first kind, respectively. These profiles satisfy the boundary conditions by virtue of the identity $J_0(z)H_1^{(1)}(z) - J_1(z)H_0^{(1)}(z) = -2i/\pi z$. Inserting this into the free energy expressions, we obtain the following expression for the total line tension of disks of diameter D:

$$\lambda^{\text{disk}}(D) = \lambda_0 - \frac{\pi D k_c \delta c_0^2}{2\sqrt{1-b^2}} \Re\Big[(\xi\alpha)^2 J_1(\alpha\frac{D}{2}) H_1^{(1)}(\alpha\frac{D}{2})\Big]. \tag{20}$$

Again, it has a minimum at finite diameter around $\sim 4\xi$.

Our result implies that domains of finite diameter become stable even before the asymptotic line tension ($D \to \infty$) vanishes: There should be a regime where the membrane is filled with small nanoscale domains, i.e., ripples or rafts. Since disk-shaped and stripe-shaped domains have similar line energies, the actual shape depends on the composition and on molecular details. We should also note that the raft domains interact with each other, hence the strong segregation theory actually predicts modulated phases with long-range order. This long-range order could be destroyed by thermal fluctuations as already discussed above in section 2.2.

2.3. Theory: Summary

To summarize, we propose a joint theoretical explanation for both ripples phases in one-component monolayers and lipid rafts in multicomponent monolayers, which is based on the coupled monolayer model. We assume that monolayers exhibit local phase separation into two phases with different order parameter (composition or other), and that the spontaneous curvature of the monolayer depends on the local order parameter. In the strong segregation limit where different phases are separated by narrow interfaces, we have shown that the line tension is reduced in the presence of a mismatch ΔK_0 between the spontaneous curvatures of the two phases. This is because monolayers with a spontaneous curvature, which are forced into being planar by the apposing monolayer, experience elastic stress. Some of that stress can be released at the domain boundaries. The resulting negative contribution to the line tension scales with $k_c\,(\delta c_0)^2$ and should be present wherever δc_0 is nonzero. A more detailed calculation shows that the elastic energy is minimized for spherical or stripe domains of a specific size, which is of the order of a few nanometers. This elastic mechanism can thus stabilize rafts of finite size for sufficiently large spontaneous curvature mismatch.

We have also considered the weak segregation limit, where the phase separation is incomplete, the interfaces are broad, and the free energy can be expanded in powers of the order parameter Φ. We have shown that the expansion has a Landau-Brazovskii form [58] (Eq. (10), with a characteristic wave vector of the order $q_0 < 1/\xi$, where $\xi \sim 1$nm is the in-plane correlation length. The correlation length ξ sets the order of magnitude and a lower limit for the characteristic wave length of phase separated structures in the membrane. The theory thus predicts local structure formation and with characteristic wave lengths of the order of a few nanometers. This is compatible with the experimental observations for ripple phases, and with experimental expectations for raft structures, should they exist in protein-free membranes. In the following section, we will show that both ripples and rafts can be observed in coarse-grained simulations, with features that support the theoretical picture.

3. Simulations

We will now describe the simulations of a generic simulation model for lipid bilayers, which reproduces both rippled states in one-component bilayers and rafts in multicomponent bilayers.

3.1. Simulation model

Lipids are represented by semiflexible linear chains of typically six tail beads attached to one slightly larger head bead [59–65]. Tail beads attract each other with truncated Lennard-Jones interactions, all other interactions are repulsive and of the Weeks-Chandler-Anderson type [66]. Lipids are immersed in a structureless fluid of so-called "phantom" solvent particles [38], which do not interact with each other (hence the fluid is structureless), only with lipids. For details of the model, we refer to Ref. [39]. The single-component model roughly reproduces the main transition and elastic properties of DPPC bilayers in the fluid phase [46], if one maps the simulation units for length and energy to SI units according to $\sigma \sim 6\AA$ and $\epsilon \sim 0.36 \cdot 10^{-20}$J [67]. In the multicomponent extension [41], "phospholipids" are mixed with "cholesterol" molecules, which are shorter and stiffer and have a slight affinity to "phospholipids" (motivated by [68]).

The model was studied by Monte Carlo simulations at constant pressure and zero membrane tension. To this end, simulations were carried out in a parallelopiped with periodic boundary conditions and all parameters of the simulation box (side lengths and angles) were allowed to fluctuate. Model lipids self-assemble spontaneously to bilayer membranes. Typical run lengths were several million Monte Carlo sweeps.

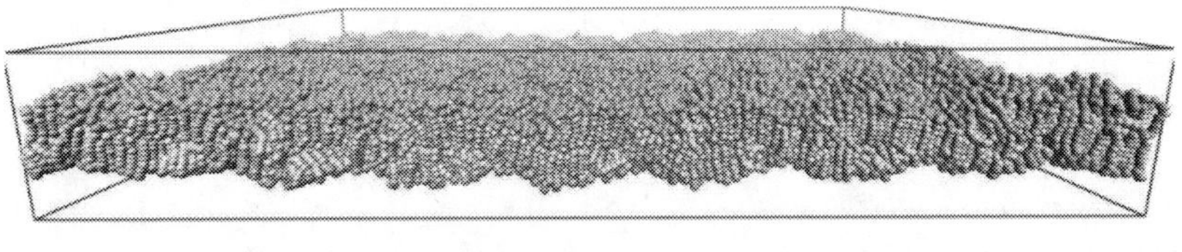

Figure 1. Asymmetric ripple state ($P_{\beta'}$ phase) in a one-component ("phospholipid") bilayer (20 000 lipids).

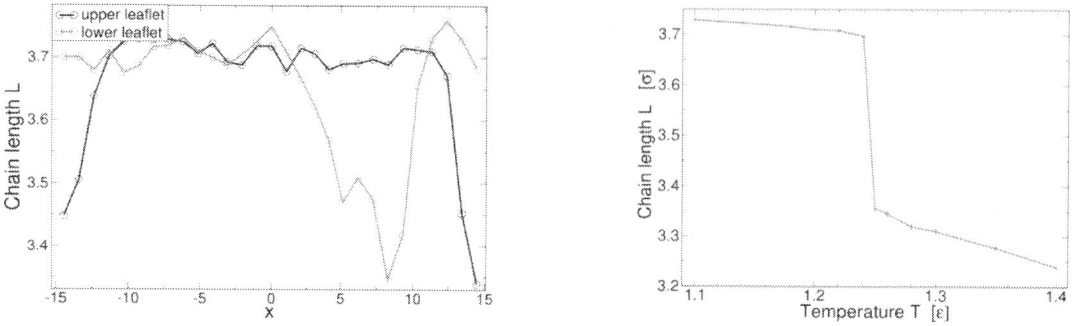

Figure 2. Left: Local chain length along the ripple direction in an asymmetric ripple state (one ripple). Right: Average chain length in the corresponding homogeneous states as a function of temperature. (At low temperature: Tilted gel phase $L_{\beta'}$; at high temperature: Fluid phase L_{α}.). From Ref. [67].

3.2. One-component membranes and ripple structures

One-component model membranes exhibit the main phases observed experimentally, i.e., the fluid phase L_α, the gel phase $L_{\beta'}$, and the ripple phase, $P_{\beta'}$. Fig. 1 shows an example of a ripple state. close to the transition to the fluid phase. One clearly recognizes how stripes with ordered lipids and curved monolayers alternate with more disordered regions. This can also be quantified, e.g., by looking at the average chain length as a function of the position in the membrane. Fig. 2 compares the profile of the average chain length in the direction perpendicular to a ripple structure (Fig. 2 a) with the average chain length in a small system undergoing a fluid-gel transition (Fig. 2 b). In the more ordered gel phase, the chains are stretched and the average chain length is roughly 10 % larger than in the disordered fluid phase. The chain length profile across the ripple structure jumps between the gel value and the fluid value. We conclude that the ripple structure can indeed be described by a sequence of locally segregated fluid and gel domains, as required by the theory of Section 2. Integrating over the difference, one finds that about one fourth of all chains are in a fluid state. This is in rough semiquantitative agreement with experimental calorimetric measurements, which suggest that about 10 % of all chains are melted [69].

In the homogeneous gel state, lipids are uniformly tilted with a tilt angle around 20 degrees. In the ripple state, the tilt angle continuously varies across the ripple structure. Fig. 3 shows the evolution across the ripple for three different ripple structures: The asymmetric ripple structure already shown in Fig. 1 (in Fig. 3 left), an intermediate structure with a single ripple defect which formed upon heating up a gel state (Fig. 3 right), and a symmetric ripple structure which is also observed occasionally both in simulations and in experiments (Fig. 3 c)). All structures are characterized by alternating gel and liquid domains in the monolayer leaflets, and in all structures, the average tilt angle varies linearly across the gel domains. Moreover, the slope of the tilt profiles is the same in all cases. ($\mathrm{d}\theta/\mathrm{d}x \sim 2.5/\sigma$ in simulation units $\sigma \sim 6\AA$). These findings underline that the ripple structures are stabilized by a propensity of lipids to splay, or in other words, by a propensity of monolayers to curve around.

3.3. Multicomponent membranes and rafts

We now turn to the discussion of multicomponent membranes. We have studied mixtures of two types of lipids, "phospholipids" and "cholesterol". These mixtures were first investigated at fixed composition (canonical ensemble) for relatively small system sizes. In particular, we have determined the chemical potential difference between the two species from the virtual acceptance

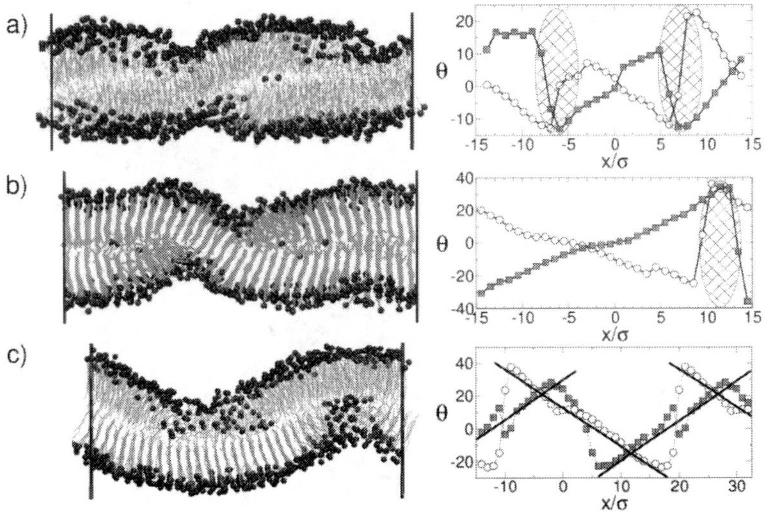

Figure 3. Configuration snapshots (left) and local tilt angle profiles along the ripple direction (right) in different ripple states. (a) Asymmetric ripple state. (b) Metastable "ripple" state with a single line defect, obtained from heating up a system from the gel state. (c) Symmetric ripple state. (from Ref. [32]).

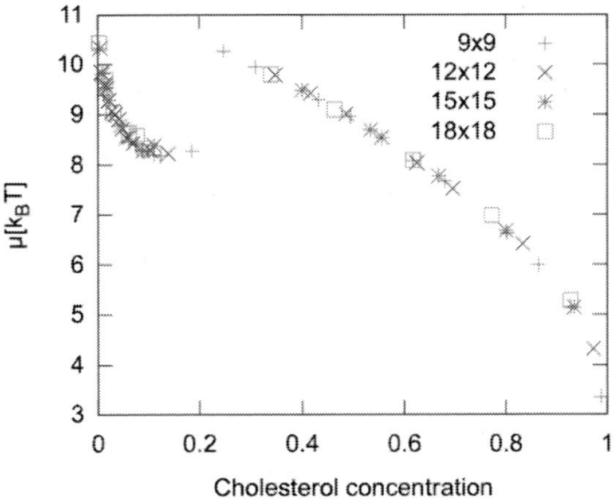

Figure 4. Chemical potential difference between "Cholesterol" and "Phospholipid" molecules in a two-component bilayer system for a range of small system sizes as indicated ($9 \times 9 \times 2$ up to $18 \times 18 \times 2$ lipids).

rate of identity switches of single molecules. (The identity switches were not performed). Fig. 4 shows the result for different system sizes as a function of cholesterol concentration for the temperature $k_B T = 1.4\epsilon$. A similar graph with a somewhat worse statistics was presented in Ref. [41]. One clearly sees the signature of a phase transition between two different phases with a transition at cholesterol concentration around 30 %. In a macroscopically stable system, the chemical potential difference must decay as a function of cholesterol concentration. Based on a Maxwell equal-area construction, one would hence expect phase separation in large systems

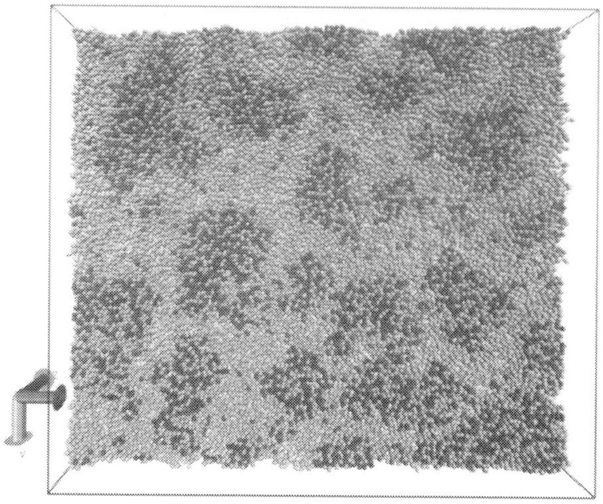

Figure 5. Rafts in a large two-component lipid bilayer (20 000 lipids). Purple (darker) beads correspond to cholesterol, green (lighter) beads to phospholipids.

between two phases with cholesterol concentrations around 5 % and 40 %.

However, the simulations of large systems tell a different story. These simulations were done in a semi-grandcanonical ensemble with constant total number of lipids, but variable composition. In case of global phase separation, the system should evolve towards one of two homogeneous phases, either an ordered phase with high cholesterol concentration, or a disordered phase with low cholesterol concentration. Instead, one obtains small ordered domains immersed in a disordered matrix: Lipid rafts (Fig. 5). Looking at the lateral structure of the membranes, we can establish that there is indeed a characteristic length scale in the system. To this end, we have identified "raft" and "non-raft" regions through a prescription based on a combination of the local density, the local composition, and the local nematic order parameter of the chains. This generated binary two-dimensional maps with black and white domains for each monolayer. The maps were then Fourier transformated. The result, radially averaged, is shown in Fig. 6 for different chemical potential differences μ. In the "unstable" region around $\mu = 8.5k_BT$, it is large, indicating strong fluctuations, and clearly features a peak at nonzero wavevector q. For higher μ, i.e., lower cholesterol concentrations, it becomes smaller and flat. Interestingly, a peak is also observed in the cholesterol-rich regime at lower $\mu = 6.2, 6.9k_BT$. This peak reflects the existence of ripple-like defects in the cholesterol-rich phase (Meinhardt et al, to be published).

We conclude that rafts of finite size with a well-defined characteristic length scale can be observed in binary mixtures. A detailed analysis of their internal structure including the stress distribution and the spontaneous curvature is under way.

One comment is in place here. Typical "raft mixtures" used for studying rafts in model membranes have at least three components. This is because three components seem necessary to bring about global lateral phase separation [14]. Here we report raft-like structures in simulations of a coarse-grained model for binary mixtures, but as in experiments, their systems do not show global phase separation between fluid states. Likewise, there is also experimental evidence that nanoscopic domains may already be present in binary mixtures – in particular mixtures of saturated lipids (lipids with high main transition temperature) and cholesterol. Studies based

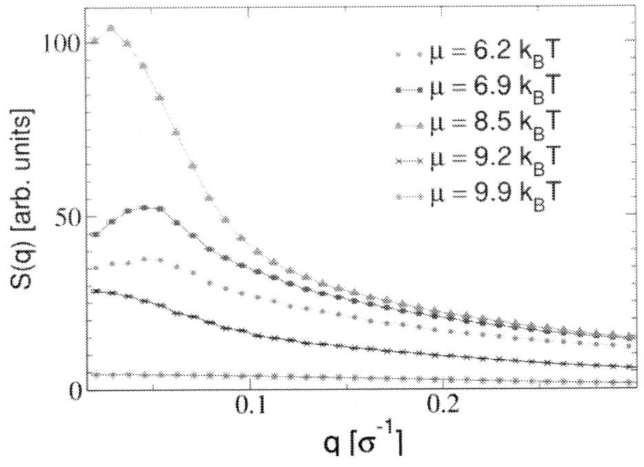

Figure 6. Radially averaged structure factor of raft conformations from semi-grandcanonical simulations of large bilayers (20 000 lipids) for different values of the chemical potential difference μ.

on local techniques such as ESR, NMR, or diffusitivy measurements have indicated the existence of immiscible liquid phases [14, 70, 71], whereas in fluorescence microscopy, one only observes one homogeneous phase [14]. This suggests that these two-component membranes phase separate on the nanoscale, while remaining homogeneous on the global scale, and that they thus feature many of the intriguing properties attributed to rafts. Shortly after we published our paper [41], this picture was supported by an experimental neutron scattering study of lipid cholesterol mixtures [72, 73].

4. Summary

In sum, we have presented a simple theory that accounts in a unified manner for both ripple phases and raft states in membranes. The prerequisites for the formation of such modulated phases is (i) local phase separation (e.g., in the ripple case, between a liquid and a gel phase, or in the raft case, between a liquid disordered and a liquid ordered phase) and (ii) curvature stress in at least one of the two phases (typically the ordered one), resulting, e.g., from a size mismatch between head group and tails. In order to reproduce rippled states or rafts, coarse-grained simulation models must meet these criteria. This is often not the case. For example, in the standard version of the popular MARTINI model [74], simulations of saturated phospholipids tend to give an untilted low-temperature gel state, and it takes very advanced sampling methods to also recover the tilted gel state [75]. Moreover, coarse-grained phospholipids and cholesterol interacting with MARTINI force fields show no sign of nonrandom mixing [76]. Therefore, it will probably be very difficult, if not impossible, to find rippled structures or nanoscale equilibrium rafts with this model.

The theory presented here is still very simplified. In future work, we will have to account for the possibility that the rafts on apposing monolayers are not rigorously coupled, and that the lipid composition on the two leaflets might be different. Moreover, the effect of fluctuations must be studied in detail in order to obtain the phase diagram, i.e., to identify parameter regions where one expects ordered modulated phases, and parameter regions where the final structure is a two dimensional microemulsion.

Our theory is generally applicable two systems that undergo local phase separation between phases with different spontaneous monolayer curvature. However, the actual coupling between the membrane composition, the order parameter, and the spontaneous curvature depends on the system under consideration. In symmetric binary membranes, it seems plausible that the order parameter is simply proportional to the difference of the concentrations of the two components. In ternary systems, the relation is more complicated, which further enriches the phase behavior. For example, binary mixtures of phospholipids and cholesterol do not demix globally above the main transition, whereas adding a third lipid component may induce lipid-lipid phase separation [14]. A possible explanation is that the third component relieves curvature stress, thus making phase separation possible. Elucidating such effects will also be an interesting direction for future research.

Acknowledgments

We have benefitted from interesting and fruitful discussions and collaborations with Dominik Düchs, Frank Brown, Jörg Neder, Michael Schick, Richard Vink, and Beate West. The simulations discussed here were carried out at the NIC computing center in Jülich and the computing centers PC2 (Paderborn) and Mogon (Mainz). This work was funded by the Deutsche Forschungsgemeinschaft within the Sonderforschungsbereich SFB 613 and SFB 625.

References

[1] Singer S J and Nicolson G K 1972 *Science* **175** 720–731
[2] Vereb G, Szöllosi J, Matko J, Nagy P, Farkas T, Vigh L, Matyus L, Waldmann T A and Damjanovich S 2003 *PNAS* **100** 8053–8058
[3] Ahmed S N, Brown D A and London E 1997 *Biochemistry* **36** 10944–10953
[4] Simons K and Ikonen E 1997 *Nature* **387** 569–572
[5] Brown D A and London E 1998 *Annu. Rev. Cell Dev. Biol.* **14** 111–136
[6] Pike L 2006 *J. Lip. Res.* **47** 1597–1598
[7] Lingwood D and Simons K 2010 *Science* **327** 46–50
[8] Leslie M 2011 *Science* **334** 1046–1047
[9] Pralle A, Keller P, Florin E L, Simons K and Hörber J K H 2000 *J. Cell. Biol.* **148** 997–1007
[10] Eggeling C, Ringemann C, Medda R, Schwarzmann G, Sandhoff K, Polyakova S, Belov V N, Hein B, von Middendorf C, Schönle A and Hell S W 2009 *Nature* **457** 1159–1162
[11] Mizuno H, Abe M, Dedecker P, Makino A, Rocha S, Ohno-Iwashita Y, Hofkens J, Kobayashi T and Miyawaki A 2011 *Chemical Science* **2** 1548–1553
[12] Owen D M, Magenau A, Williamson D and Gaus K 2012 *Bioessays* **34** 739–747
[13] Veatch S L and Keller S L 2003 *Biophys. J.* **85** 3074–3083
[14] Veatch S L and Keller S L 2005 *Biochimica et Biophysica Acta* **1746** 172–185
[15] Veatch S L and Keller S L 2005 *Phys. Rev. Lett.* **94** 148101
[16] Veatch S L, Cicuta P, Sengupta P, Honerkamp-Smith A, Holowka D and Baird B 2008 *ACS Chem. Biol.* **3** 287–293
[17] Turner M S, Sens P and Socci N D 2005 *Phys. Rev. Lett.* **95** 168301
[18] Kuzmin P I, Akimov S A, Chizmadzhev Y A, Zimmerberg J and Cohen F S 2005 *Biophys. J.* **88** 1120–1133
[19] Yethiraj A and Weisshaar J C 2007 *Biophys. J.* **93** 3113–3119
[20] Simons K and Vaz W L C 2004 *Annu. Rev. Biophys. Biomol. Struct.* **33** 269–295
[21] Brewster R and Pincus P A Safran S A 2009 *Biophys. J.* **97** 1087–1094
[22] Yamamoto T, Brewster R and Safran S A 2010 *EPL* **91** 28002
[23] Hirose Y, Komura S and Andelman D 2012 *Phys. Rev. E* **86** 021916
[24] Schick M 2012 *Phys. Rev. E* **85** 031902
[25] Veatch S L, Soubias O, Keller S L and Gawrisch K 2007 *PNAS* **104** 17650–17655
[26] Honerkamp-Smith A R, Veatch S L and Keller S L 2009 *Biochimica et Biophysica Acta* **1788** 53–63
[27] Sengupta K, Raghunathan V A and Katsaras J 2003 *Phys. Rev. E* **68** 031710
[28] Koynova R and Caffrey M 1994 *Chemistry and Physics of Lipids* **69** 1–34
[29] Koynova R and Caffrey M 1998 *Biochimica et Biophysica Acta - Reviews on Biomembranes* **1376** 91–145
[30] Kranenburg M and Smit B 2005 *J. Phys. Chem. B* **109** 6553–6563
[31] de Vries A H, Yefimov S, Mark A E and Marrink S J 2005 *PNAS* **102** 5392–5396

[32] Lenz O and Schmid F 2007 *Phys. Rev. Lett.* **98** 058104
[33] Sun X and Gezelter J D 2008 *J. Phys. Chem. B* **112** 1968–1975
[34] Jamroz D, Kepczynski M and Nowakowska M 2010 *Langmuir* **26** 15076–15079
[35] Chen R, Poger D and Mark A E 2011 *J. Phys. Chem. B* **115** 1038–1044
[36] Dolezel S 2010 *Computer simulation of lipid bilayers* Diploma thesis Universität Mainz
[37] Düchs D and Schmid F 2001 *J. Phys.: Cond. Matt.* **13** 4853
[38] Lenz O and Schmid F 2005 *J. Mol. Liquids* **117** 147–152
[39] Schmid F, Düchs D, Lenz O and West B 2007 *Comp. Phys. Comm.* **177** 168
[40] West B and Schmid F 2010 *Soft Matter* **6** 1275–1280
[41] Meinhardt S, Vink R and Schmid F 2013 *PNAS* **12** 4476–4481
[42] Dan N, Pincus P and Safran S A 1993 *Langmuir* **9** 2768–2771
[43] Dan N, Berman A, Pincus P and Safran S A 1994 *J. de Physique II* **4** 1713
[44] Aranda-Espinoza H, Berman A, Dan N, Pincus P and Safran S 1996 *Biophys. J.* **71** 648
[45] Brannigan G and Brown F 2006 *Biophys. J.* **90** 1501–1520
[46] West B, Brown F L H and Schmid F 2009 *Biophys. J.* **96** 101–115
[47] Collins M D and Keller S L 2008 *PNAS* **105** 124–128
[48] Safran S A 1994 *Statistical Thermodynamics of Surfaces, Interfaces, and Membranes* (Cambridge, Massachusetts: Perseus Books)
[49] Neder J, West B, Nielaba P and Schmid F 2010 *J. Chem. Phys.* **132** 115101
[50] Akimov S A, Kuzmin P I, Zimmerberg J and Cohen F S 2005 *Phys. Rev. E* **75** 011919
[51] Watson M C, Morriss-Andrews A, Welch P M and Brown F L H 2013 *JCP* **139** 084706
[52] Neder J, Nielaba P, West B and Schmid F 2012 *New J. of Physics* **14** 125017
[53] Leibler S and Andelman D 1987 *J. de Physique* **48** 2013–2018
[54] Kollmitzer B, Heftberger P, Rappolt M and Pabst G 2013 *Soft Matter* **preprint (accepted)**
[55] Loison C, Mareschal M, Kremer K and Schmid F 2003 *J. Chem. Phys.* **119** 13138–13148
[56] Loison C, Mareschal M and Schmid F 2004 *J. Chem. Phys.* **121** 1890–1900
[57] Rawicz W, Olbrich K C, McIntosh T, Needham D and Evans E 2000 *Biophys. J.* **79** 328–339
[58] Brazovskii S A 1975 *Soviet Physics JETP* **41** 85–89
[59] Haas F, Hilfer R and Binder K 1995 *J. Chem. Phys.* **102** 2960–2969
[60] Haas F, Hilfer R and Binder K 1996 *J. Phys. Chem.* **100** 15290–15300
[61] Haas F and Hilfer R 1996 *J. Chem. Phys.* **105** 3859–3867
[62] Stadler C, Lange H and Schmid F 1999 *Phys. Rev. E* **59** 4248
[63] Stadler C and Schmid F 1999 *J. Chem. Phys.* **110** 9697
[64] Schmid F, Stadler C and Düchs D 2001 *J. Phys.: Cond. Matt.* **13** 8653–8659
[65] Schmid F 2009 *Macromol. Rap. Comm.* **30** 741–751
[66] Weeks J D, Chandler J and Andersen H C 1971 *J. Chem. Phys.* **54** 5237–5247
[67] Lenz O 2007 *Computer simulations of lipid bilayers* Dissertation Universität Bielefeld
[68] Lindblom G and Orädd G 2009 *Biochimica et Biophysica Acta* **1788** 234–244
[69] Nagle J F 1980 *Annu. Rev. Phys. Chem.* **31** 157–195
[70] Ipsen J H, Karlström G, Mouritsen O B, Wennerström H and Zuckermann M J 1987 *Biochimica et Biophysica Acta* **905** 162–172
[71] Sankaram M B and Thompson T E 1991 *PNAS* **88** 8686–8690
[72] Armstrong C L, Marquardt D, Dies H, Kucerka N, Yamani Z, Harroun T A, Katsaras J, Shi A C and Rheinstädter M C 2013 *PLoS One* **8** E66162
[73] Rheinstädter M C and Mouritsen O G 2013 *Current Opinion in Colloid & Interface Science* **preprint**
[74] Marrink S J, Risselada H J, Yefimov S, Tielemann D P and de Vries A H 2007 *J. Phys. Chem. B* **111** 7812–7823
[75] Nagai T, Ueoka R and Okamoto Y 2012 *J. Phys. Soc. Jpn.* **81** 024002
[76] Waheed Q, Tjörnhammar R and Edholm O 2012 *Biophys. J.* **103** 2125–2133

Journal of Physics: Conference Series **487** (2014) 012005 doi:10.1088/1742-6596/487/1/012005

Bicritical or tetracritical: the 3D anisotropic Heisenberg antiferromagnet

Shan-Ho Tsai[1,2], Siyan Hu[1], and D. P. Landau[1]

[1] Center for Simulational Physics, University of Georgia, Athens, GA 30602, USA
[2] Georgia Advanced Computing Resource Center, Enterprise Information Technology Services, University of Georgia, Athens, GA 30602, USA

E-mail: shtsai@uga.edu

Abstract.
The classical uniaxially anisotropic Heisenberg antiferromagnet on the simple cubic lattice, in the presence of an external magnetic field, is believed to have a multicritical point; however, there has been controversy whether it is a bicritical or a tetracritical point. We perform Monte Carlo simulations of this model and analyze the components of the staggered magnetization, the susceptibilities and the probability distribution of the magnetization to conclude that the multicritical point is bicritical and it is in the three-dimensional Heisenberg universality class.

1. Introduction

The Hamiltonian of the uniaxially anisotropic Heisenberg antiferromagnetic model is given by:

$$\mathcal{H} = J \sum_{\langle i,j \rangle} [\Delta(S_i^x S_j^x + S_i^y S_j^y) + S_i^z S_j^z] - H \sum_{i=1}^{N} S_i^z \tag{1}$$

where the three-component unit vectors $\mathbf{S}_i = (S_i^x, S_i^y, S_i^z)$ represent classical spins on sites i of a simple cubic lattice, with linear size L, and $J > 0$ is the exchange coupling between nearest-neighbor pairs of spins. The first summation is over all $\langle i,j \rangle$ pairs of neighboring sites and the second summation is over all $N = L^3$ lattice sites. The uniaxial exchange anisotropy Δ is set to $\Delta = 0.8$, which gives rise to an easy axis in the z direction. A magnetic field H is applied along the $z-$axis. This model is also referred to as the XXZ antiferromagnet and it provides a good description of many real magnetic systems, including $NiCl_26H_2O$ [1, 2, 3], MnF_2 [4, 5, 6], and $GdAlO_3$ [7, 8, 9], among others. A different anisotropic Heisenberg model, which has a crystal field term, is studied in Ref.[10].

Early leading-order renormalization group theory [11], Monte Carlo simulations [12], high-temperature series expansion [13], among other studies, concluded that the model described above has an antiferromagnetic (AF) phase (see Fig. 1(a)) at low temperature T and low field H, a spin-flop (SF) phase (see Fig. 1(b)) at low T and higher H, and a paramagnetic (P) phase at high T and/or at high H. The AF to P phase transition line is in the Ising universality class, while the SF to P phase transition line is in the XY universality class in this picture. The AF and SF phases are separated by a first-order phase transition line and the three phases meet at a bicritical point, which is in the three-dimensional (3D) Heisenberg universality class. Later, a renormalization group theory in higher-loop order [14] finds the presence of a tetracritical point for this model with a new biconical phase (see Fig. 1(c)). A subsequent renormalization group study in two-loop order proposes that a bicritical point in the 3D Heisenberg universality class cannot be excluded [15]. Recent Monte Carlo simulations using Metropolis sampling, focusing on simple cubic lattices with linear sizes L up to 32, and carrying out critical property analysis using finite size scaling on the multicritical point [16, 17] corroborate a scenario with a bicritical point in the 3D Heisenberg universality class. However, a subsequent numerical analysis of anisotropic perturbations in three-dimensional O(N)-symmetric vector models suggests [18] that for the model in question here, the stable fixed point has a biconical structure (see Fig. 1(c)), and its critical exponents are very close to the ones in the 3D Heisenberg universality class [19]. Therefore, it is very hard to distinguish between a bicritical point in the Heisenberg universality class and a tetracritical point (with a biconical phase) using Monte Carlo simulations and finite-size scaling to determine critical exponents at the multicritical point. Figure 2 shows a schematic view of the different phase diagrams that have been proposed for this model.

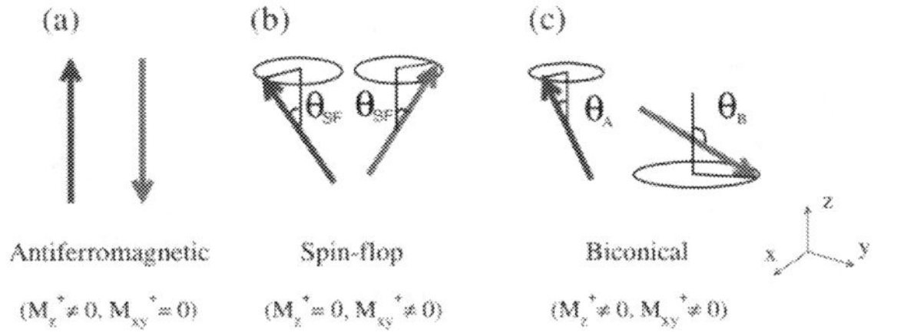

Figure 1. Illustration of spin configurations of the sublattices in the antiferromagnetic, spin-flop, and biconical phases.

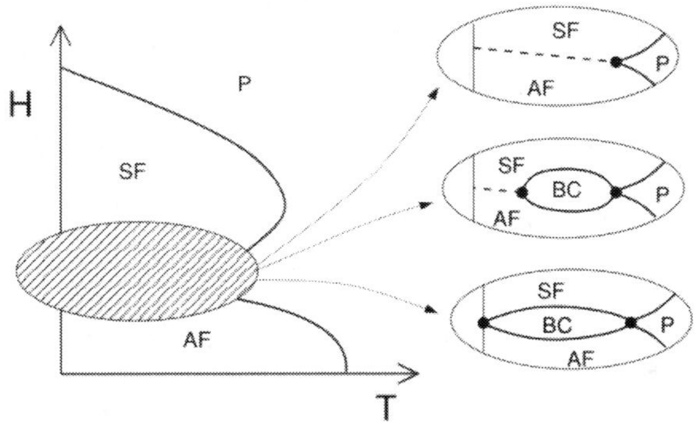

Figure 2. Schematic plot of the phase diagram with the three suggested scenarios for the region near the multicritical point. AF, SF, BC, and P label an antiferromagnetic, a spin-flop, a biconical, and a paramagnetic phase, respectively.

In this paper we use Monte Carlo simulations with a hybrid sampling method that includes Metropolis and Wolff-cluster steps, and use larger lattices to determine the nature of the multicritical point. Instead of obtaining critical exponents at the multicritical point, we examine the order of the AF to SF phase transition below, but near, the multicritical point and investigate whether a biconical phase is present. We also compare the probability distribution of the uniform magnetization, which is a universal quantity, with the distribution of the 3D Heisenberg universality class.

2. Simulation methods

We carried out Monte Carlo simulations [20] for the model described in Eq.(1) using a hybrid method that consists of Metropolis single-spin-flip sampling [21], combined with the Wolff-cluster algorithm [22] applied to the z-component of the spins, i.e. S_i^z. A typical run consists of $10^7 \sim 1.5 \times 10^8$ hybrid Monte Carlo steps (MCS), where each hybrid Monte Carlo step consists of 6 Wolff-cluster steps and 4 sweeps through the lattice using Metropolis sampling. The Wolff-cluster algorithm implemented here satisfies detailed balance, but it is not ergodic, because it only changes the z component of the spins ; however, the hybrid sampling method used here is ergodic and it satisfies detailed balance. This hybrid sampling method allowed us to use larger lattice sizes, with linear dimensions $10 \le L \le 60$. Periodic boundary conditions are used to minimize finite size effects. We also use histogram reweighting methods [23] to obtain results for temperatures and external fields near the values at which simulations were performed. Several independent runs were carried out to compute error bars in the quantities; whenever error bars are not shown in the figures, they are of the size of or smaller than the symbol sizes.

In order to determine the phase transition line between the AF and SF phases and to detect a possible biconical phase, we examine the components of the staggered magnetization along the z direction and on the xy plane, *i.e.* M_z^+ and M_{xy}^+, respectively. These quantities are defined by

$$M_z^+ = |M_{z,A} - M_{z,B}| \tag{2}$$

$$M_{xy}^+ = \sqrt{(M_{x,A} - M_{x,B})^2 + (M_{y,A} - M_{y,B})^2} \tag{3}$$

where $(M_{x,i}, M_{y,i}, M_{z,i})$ are the x, y, and z components of the uniform magnetization on sublattices $i = A, B$. In the AF phase, M_z^+ is non-zero and M_{xy}^+ is zero, whereas in the SF phase M_z^+ vanishes but M_{xy}^+ does not. Therefore these two quantities can be used as order parameters for the phase transitions between the AF and SF phases. A concurrent, non-zero value of M_z^+ and M_{xy}^+ would reveal a biconical phase. The susceptibilities χ_{xy}^+ and χ_z^+ are computed from the fluctuations of M_{xy}^+ and M_z^+, respectively. To simplify notation, we set $J = 1$ and the Boltzmann constant $k_B = 1$.

3. Results

The phase diagram for the XXZ model described by the Hamiltonian in Eq. (1), as obtained by Selke [17], is shown in Fig. 1 of Ref.[17]. The multicritical point where the AF, SF, and P phases meet was determined to be [17] at $k_BT/J = 1.025 \pm 0.0025$ and $H/J = 3.89 \pm 0.01$.

Below the multicritical point, if we fix the temperature T and vary the external field H from the AF to the SF phase, we would see a single first-order phase transition point in the absence of a biconical phase. If a biconical phase exists, then two second-order phase transition points, separating the AF and BC phases and the BC and SF phases, would be seen.

At $T = 0.95$ the ensemble average of the order parameters M_{xy}^+ and M_z^+ per site as a function of H, shown in Fig. 3, indicate that a single phase transition occurs around $H = 3.8382$ (shown by the dashed line in Fig. 3). The peak positions of the susceptibilities χ_{xy}^+ and χ_z^+ (see Fig. 4) occur at the same field (marked by dashed line in Fig. 4), which corroborates that a single phase transition between the AF and the SF phase occurs at this temperature. Hence, we see no evidence of a biconical phase at $T = 0.95$.

Next, we consider a temperature closer to the multicritical point, namely $T = 0.98$. Fig. 5 shows the ensemble average of the order parameters M_{xy}^+ and M_z^+ per site as a function of H at $T = 0.98$. Again, the curves for the larger lattice sizes show that a single phase transition happens around $H = 3.8614$ (shown by the dashed line in Fig. 5). The presence of a single phase transition is confirmed by the location of the peaks in χ_{xy}^+ and χ_z^+, which occur at the same field (shown by the dashed line in Fig. 6). We have also plotted the order parameters and their susceptibilities at $T = 1.01$ (results not shown here), and we have seen no evidence of a biconical phase.

The order of the AF to SF phase transition points at temperatures below the multicritical point is determined with a finite-size scaling plot of the logarithm of the maximum of the susceptibilities χ_{xy}^+ and χ_z^+ as a function of L. For a first-order phase transition, the maximum

Journal of Physics: Conference Series **487** (2014) 012005 doi:10.1088/1742-6596/487/1/012005

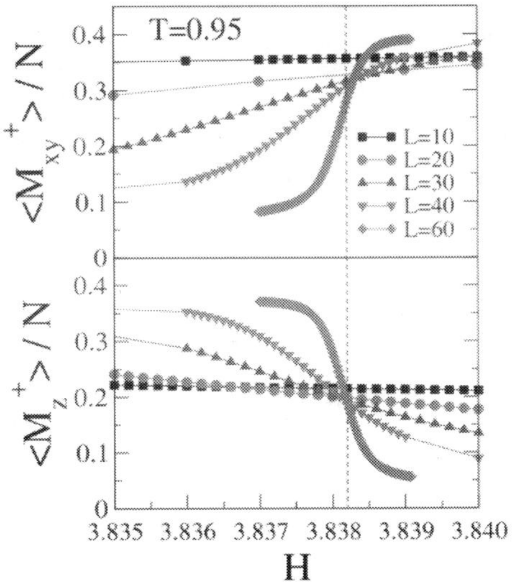

Figure 3. Components of the staggered magnetization per site as a function of H for $T = 0.95$. The dashed line marks $H = 3.8382$.

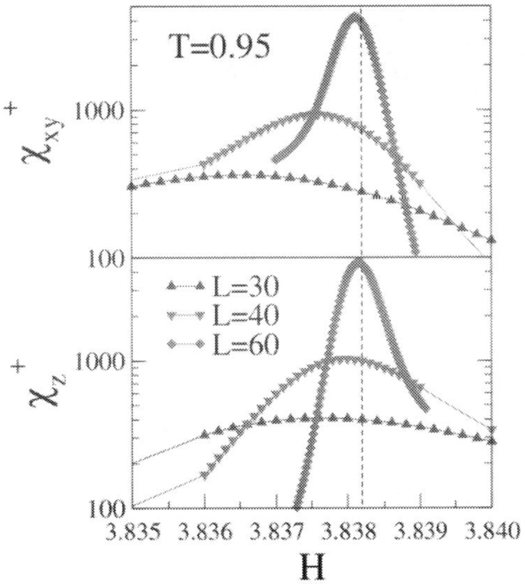

Figure 4. Magnetic susceptibilities χ^+_{xy} and χ^+_z as a function of the external field H for $T = 0.95$. The dashed line marks $H = 3.8382$.

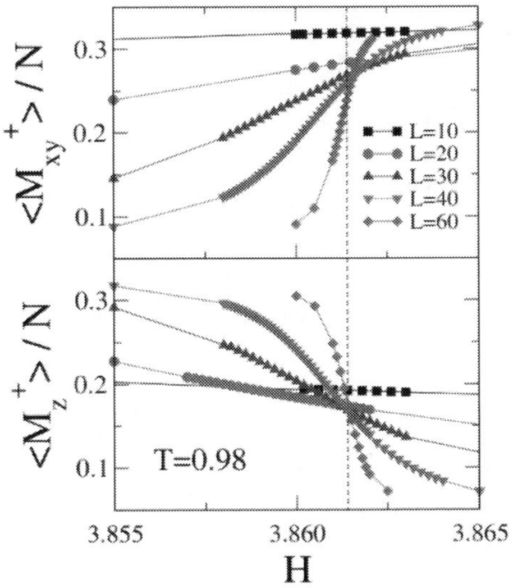

Figure 5. Components of the staggered magnetization per site as a function of H for $T = 0.98$. The dashed line marks $H = 3.8614$.

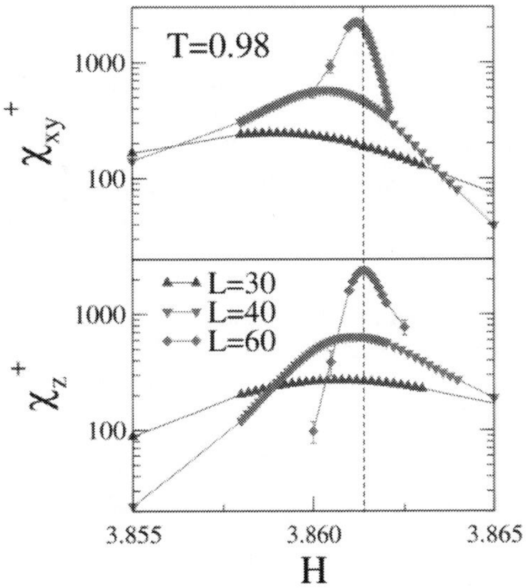

Figure 6. Magnetic susceptibilities χ^+_{xy} and χ^+_z as a function of external field H for $T = 0.98$. The dashed line marks $H = 3.8614$.

of the susceptibility of the order parameter scales as L^d, where d is the dimension of the lattice. Linear least-squares fitting of the maximum of χ^+_{xy} and χ^+_z as a function of L in a log-log plot (see Figs. 7(a) and 7(b)) show that the slopes in these graphs are very close to 3, which is the dimension of the simple cubic lattice used here. For $T = 1.01$, due to the proximity to the multicritical point, the linear fitting gives a slope close to 3 when the smaller lattices ($L = 10, 20$) are excluded from the fitting. These results indicate that the AF to SF phase transition points at these temperatures below, but quite near, the multicritical point are of first order.

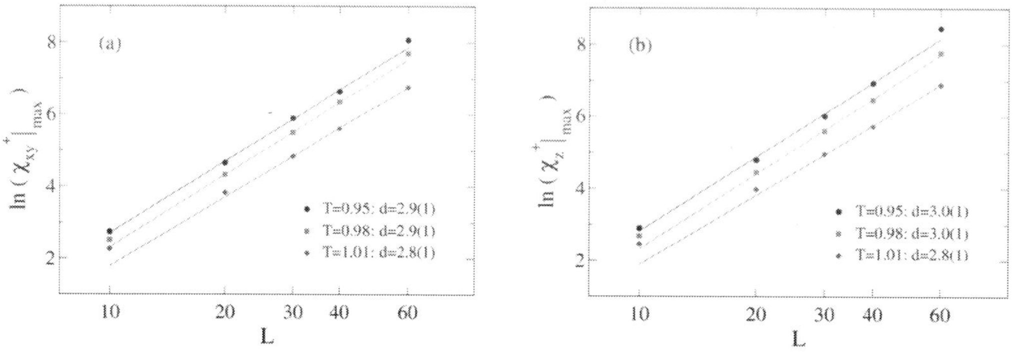

Figure 7. Logarithm of the maximum of the susceptibilities (a) χ^+_{xy} and (b) χ^+_z as a function of L for temperatures below the multicritical point. The linear fitting for $T = 0.95$, $T = 0.98$, and $T = 1.01$ includes data points for $L = 10$ to 60, $L = 20$ to 60, and $L = 30$ to 60, respectively.

In order to determine the nature of the multicritical point, we consider the universal scaling function $P^*(m/\sigma)$, where σ is the variance of the uniform magnetization $m = \sqrt{M_x^2 + M_y^2 + M_z^2}$, which we rescale to m/σ. The probability distribution has the following form [25, 26]:

$$P(m) = \frac{1}{\sigma} P^*(m/\sigma) \tag{4}$$

where σ is calculated as $\sigma^2 = <m^2> - <m>^2$. Fig. 8 shows the comparison of our probability distribution for $L = 30$ and $L = 40$ with the probability distribution for the $3D$ Heisenberg model. We used histogram reweighting in T and H to obtain curves that best match the $3D$ Heisenberg probability distribution. The distribution for the Heisenberg model was obtained for $L = 40$, at the critical point [24]. These unit-variance, zero-average curves collapse very nicely; hence, we have confidence in concluding that the meeting point of the AF, SF, and P phases belongs to the $3D$ Heisenberg universality class.

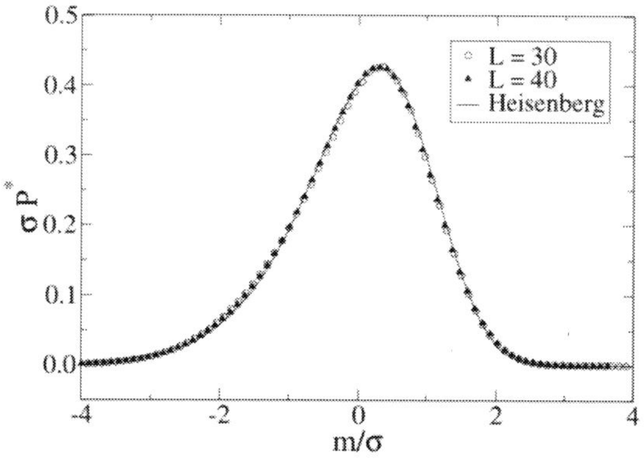

Figure 8. Scaling function $P^*(m/\sigma)$ for $L = 30$ and $L = 40$ at the multicritical point, and for the 3D Heisenberg model at its critical point.

4. Conclusions

Monte Carlo simulations with Wolff-cluster and Metropolis sampling methods are used to study the uniaxially anisotropic Heisenberg antiferromagnet (XXZ model) in an external field H along the z direction, on simple cubic lattices. By looking at fixed temperatures $T = 0.95$, $T = 0.98$, and $T = 1.01$ below the multicritical point, we determined that for each T there is a single first-order phase transition between the antiferromagnetic and the spin-flop phases. We have seen no evidence of a biconical phase. Because the critical exponents at a tetracritical point are very close to those at a bicritical point in the 3D Heisenberg universality class, it is very hard to use critical exponent values to determine the nature of the multicritical point. Hence, we used the probability distribution of the magnetization for this purpose. An excellent collapse of this probability distribution with the universal distribution of the $3D$ Heisenberg model leads us to conclude that the multicritical point in the XXZ model is indeed a bicritical point in the $3D$ Heisenberg universality class.

Acknowledgments

We are grateful to J. A. Plascak for very helpful discussions. This work was supported by NSF Grant number DMR-0810223 and it was supported in part by resources from the Georgia Advanced Computing Resource Center, a partnership between the University of Georgia's Office of the Vice President for Research and Office of the Vice President for Information Technology.

References

[1] Oliveira Jr N F, Paduan Filho A, Salinas S R 1975 *Phys. Lett.* A **55** 293
[2] Oliveira Jr N F, Paduan Filho A, Salinas S R 1976 *AIP Conf. Proc.* **29** 463
[3] Oliveira Jr N F, Paduan Filho A, Salinas S R and Becerra C C 1978 *Phys. Rev.* B **18** 6165
[4] Shapira Y and Foner S 1970 *Phys. Rev.* B **1** 3083
[5] King A R and Rohrer H 1976 *AIP Conf. Proc.* **29** 420
[6] Shapira Y and Becerra C C 1976 *Phys. Lett.* A **57** 483
[7] Rohrer H 1975 *Phys. Rev. Lett.* **34** 1638
[8] Rohrer H 1975 *AIP Conf. Proc.* **24** 268
[9] Rohrer H and Gerber Ch 1977 *Phys. Rev. Lett.* **38** 909
[10] Freire R T S, Martins P H L and Plascak J A *Journal of Physics Conference Series* current volume
[11] Nelson D R, Kosterlitz J M and Fisher M E 1974 *Phys. Rev. Lett.* **33** 813
[12] Landau D P and Binder K 1978 *Phys. Rev.* B **17** 2328
[13] Mouritsen O G, Kjaersgaard Hansen E and Knak Jensen S J 1980 *Phys. Rev.* B **22** 3256
[14] Calabrese P, Pelissetto A and Vicari E 2003 *Phys. Rev.* B **67** 054505
[15] Folk R, Holovatch Y and Moser G 2008 *Phys. Rev.* E **78** 041124
[16] Bannasch G and Selke W 2009 *Euro. Phys. J.* B **69** 439
[17] Selke W 2011 *Phys. Rev.* E **83** 042102
[18] Hasenbusch M and Vicari E 2011 *Phys. Rev.* B **84** 125136
[19] Campostrini M, Hasenbusch M, Pelissetto A, Rossi P and Vicari E 2002 *Phys. Rev.* B **65** 144520
[20] Landau D P and Binder K 2009 *A Guide to Monte Carlo Simulations in Statistical Physics* (Cambridge: Cambridge University Press)
[21] Metropolis N, Rosenbluth A W, Rosenbluth M N, Teller A H and Teller E 1953 *J. Chem. Phys.* **21** 1087
[22] Wolff U 1989 *Phys. Rev. Lett.* **62** 361
[23] Ferrenberg A M and Swendsen R H 1988 *Phys. Rev. Lett.* **61** 2635
[24] Chen K, Ferrenberg A M and Landau D P 1993 *Phys. Rev.* B **48** 3249
[25] Martins P H L and Plascak J A 2007 *Phys. Rev.* E **76** 012102
[26] Plascak J A and Martins P H L 2013 *Comp. Phys. Comm.* **184** 259

Journal of Physics: Conference Series 487 (2014) 012006 doi:10.1088/1742-6596/487/1/012006

Location of the Bicritical Point of the Anisotropic Heisenberg Model in a Crystal Field

R T S Freire[1*], **P H L Martins**[2†], **J A Plascak**[3,4‡]

[1] Universidade Federal de São João del Rei, Campus Dom Bosco
Praça Dom Helvécio, 74 - Fábricas
36301-160 - São João del Rei, MG - Brazil
[2] Instituto de Física, Universidade Federal de Mato Grosso,
Av. Fernando Corrêa da Costa, 2367, 78060-900, Cuiabá, MT - Brazil
[3] Departamento de Física, Instituto de Ciências Exatas,
Universidade Federal de Minas Gerais, C.P. 702, 30123-970, Belo Horizonte, MG - Brazil
[4] Center for Simulational Physics, University of Georgia, Athens, GA 30602, USA

E-mail: * rfreire@ufsj.edu.br, † pmartins@fisica.ufmt.br, ‡ pla@fisica.ufmg.br

Abstract. Extensive Monte Carlo simulations are used to locate the bicritical point of the ferromagnetic anisotropic Heisenberg model on a simple cubic lattice in the presence of a crystal field. In order to achieve this goal, we employ a hybrid algorithm comprising Metropolis and Wolff algorithms along with field mixing. Multiple histogram techniques have also been used to obtain the probability distribution function of the conjugate extensive variables of interest. In conjunction with finite-size scaling analysis we computed the fourth-order cumulant of the order parameter along the first-order line and the corresponding bicritical point has been located from the crossings of the cumulants.

1. INTRODUCTION

In a recent paper [1], Plascak and Martins review the importance of the order parameter probability distribution to investigate the critical properties and locate multicritical points through Monte Carlo simulations in statistical mechanics models, in particular discrete state models. This tool has been successfully employed along with field mixing techniques to locate multicritical points, for example the double critical end point of the two-dimensional spin-3/2 Blume-Capel model [2], the tricritical point of a two-dimensional spin fluid [3], the asymmetric Ising model [4], the two-dimensional Potts model [5], the three-dimensional XXZ model [6], among others. In particular, we have already used this technique to locate the tricritical point of the XY vector Blume-Emery-Griffiths model in three dimensions [7]. Here we once again apply this technique to study a continuous spin model.

In a previous work [8] we have investigated, through Monte Carlo simulations [9], the phase diagram of the anisotropic ferromagnetic Heisenberg model in a crystal field on a simple cubic lattice, which reveals a rich topology due to the competition between crystalline and exchange anisotropies. This phase diagram, depicted in Fig. 1 for a finite lattice of linear size $L = 14$, shows two ordered phases, separated by a first-order transition line. The Ising-like ordered phase has an easy axis ordering (z axis), while the XY-like phase has an easy plane ordering, perpendicular to the z axis. Each ordered phase is separated from the disordered paramagnetic

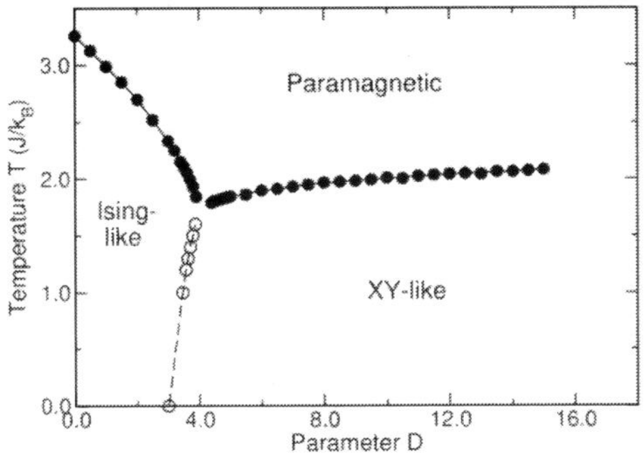

Figure 1. Phase diagram for the anisotropic Heisenberg model in a crystal field on a simple cubic lattice (L=14). Lines are only guide to the eyes. The continuous lines and full circles indicate second-order phase transitions, while the dashed line and open circles stand for first order phase transitions. As mentioned in [8], the finite-size scaling behavior performed revealed that the extrapolated temperatures were very close (less than 1% discrepant) to those obtained by taking L=14, justifying, for our purposes, the use of such lattice size to obtain the phase diagram. Error bars are of the size or smaller than symbol sizes.

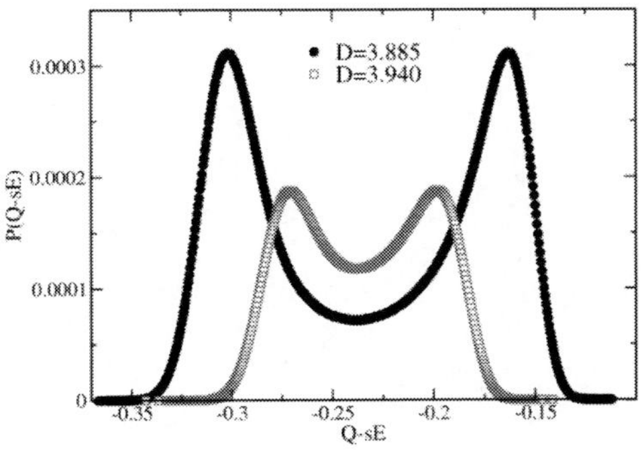

Figure 2. Probability distribution function for the field mixing extensive conjugate variable $\mathcal{Q} = Q - sE$ (L=20). The distribution displayed refers to two different points along the first-order line. Error bars are of the size or smaller than symbol sizes.

phase by a second-order transition line. All transition lines terminate at a bicritical point. It is the location of this bicritical point that we want now to determine with better precision.

The model is defined by the hamiltonian

$$\mathcal{H} = -J \sum_{\langle i,j \rangle} (\vec{S}_i \cdot \vec{S}_j) - A \sum_{\langle i,j \rangle} S_i^z S_j^z + D \sum_{i=1}^{N} (S_i^z)^2, \tag{1}$$

where $\vec{S}$ stands for a classical three dimensional spin vector with unity modulus and $\langle i,j \rangle$ stands for a sum over first-neighbor spins on an N spin simple cubic lattice. J is the ferromagnetic ($J = 1$) exchange interaction parameter, A refers to the easy-axis exchange anisotropy ($A > 0$) and D is related to the easy-plane crystalline interaction (we consider herein $D > 0$). In this work we used $A = 1$.

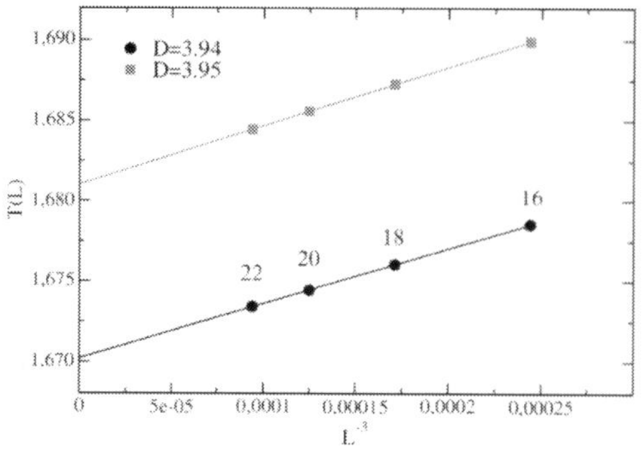

Figure 3. Finite size scaling analysis for two different points along the first-order line. The line is the best fit to the data points for the indicated lattice sizes. Error bars are smaller than symbol sizes.

2. RESULTS

Simulations were performed on a simple cubic lattice with L=16, 18, 20 and 22. In order to reduce critical slowing down, we used a hybrid Monte Carlo scheme [10], performing four Wolff [11] steps after one Metropolis [12] sweep. The Wolff algorithm was adapted to take into account the anisotropies and dimensionality of the model based on the scheme proposed in Ref. [13]. In order to achieve good precision we had to perform very long runs (ranging from 280 to 690 million hybrid Monte Carlo steps).

Using field mixture [1, 14] and multiple histogram techniques [9, 15, 16, 17], we probed the probability distribution of the conjugate extensive variable associated with the parameters D and T in order to precisely locate the first-order transition line. This was achieved tuning the values of the mixing parameter s and temperature T for a given value of the parameter D and a specific lattice size L in order to obtain a symmetric distribution, as shown in Fig. 2 for two values of the parameter D. Having the finite-size pseudo-transition temperature for each lattice size, we performed the finite-size scaling analysis and determined the transition temperature. In Fig. 3, the finite size scaling analysis is shown for two different values of the parameter D along

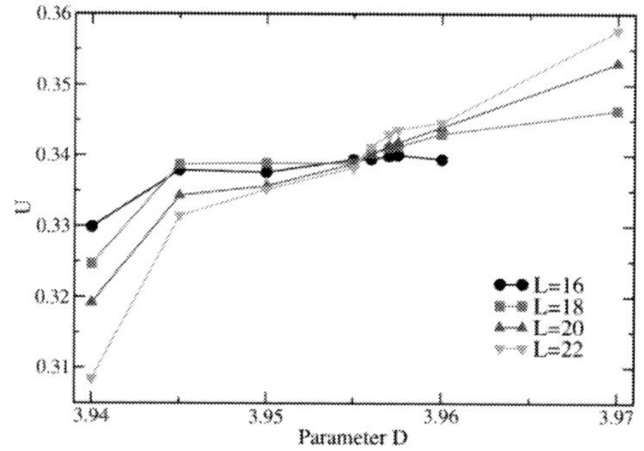

Figure 4. Fourth-order cumulant of the order parameter for the three-dimensional anisotropic Heisenberg model. Lines are only guide to the eyes and the error bars are smaller than the symbol sizes.

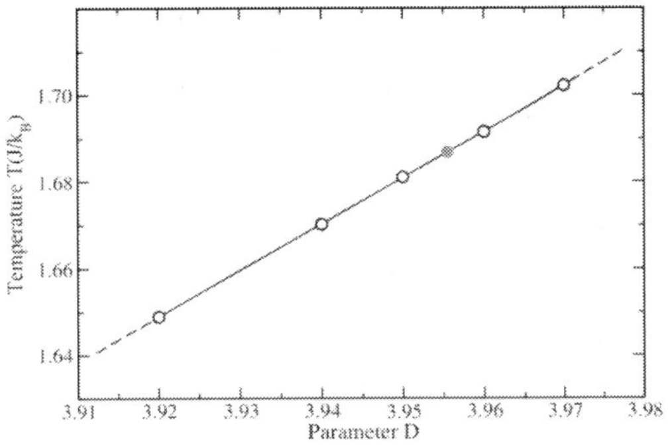

Figure 5. First-order transition line and its analytical extension on the $D-T$ plane, obtained with the procedure described in the text. The bicritical point is depicted in red. Lines are only guide to the eyes and the error bars are smaller than the symbol sizes.

the first-order line. The conjugate extensive variable of interest here is $\mathcal{Q} = Q - sE$, where

$$Q = \sum_{i=1}^{N} (S_i^z)^2, \tag{2}$$

and

$$E = \sum_{\langle i,j \rangle} (\vec{S}_i \cdot \vec{S}_j). \tag{3}$$

We then computed the fourth-order cumulant of the order parameter (the z-component of the magnetization) along the first-order line, defined as

$$U = 1 - \frac{\langle m_z^4 \rangle}{3 \langle m_z^2 \rangle^2}, \tag{4}$$

where

$$m_z = \sum_{i=1}^{N} S_i^z. \tag{5}$$

At the bicritical point, the fourth-order cumulant should coincide for sufficiently large system sizes. Thus, following the first-order transition line, we searched the cumulant crossing in order to determine the bicritical point.

According to the cumulant crossing (Fig. 4), the bicritical point was located at $D = 3.9555(4)$, which corresponds to a bicritical temperature $T = 1.6868(4)$. In Fig. 5, the first-order transition line is depicted, along with its analytical extension, whose beginning is marked off by the bicritical point.

3. CONCLUSIONS

In summary, employing field mixing and multiple histogram techniques, we have located the bicritical point of the three-dimensional anisotropic Heisenberg model with very good precision at $D = 3.9555(4)$ and $T = 1.6868(4)$. Further results concerning the bicritical behavior of this model will be published elsewhere.

4. ACKNOWLEDGMENTS

RTSF thanks Prof. Gerald Weber for generous allocation of CPU time and disk space at the Statistical Physics Simulation Laboratory, Universidade Federal de Minas Gerais. Financial support from CNPq, CAPES and FAPEMIG are gratefully acknowledged.

5. REFERENCES

[1] Plascak, J A and Martins, P H L 2013 *Computer Physics Communications* **184**(2) 259
[2] Plascak, J A and Landau, D P 2003 *Phys. Rev. E* **67** 015103(R)
[3] Wilding, N B and Nielaba, P 1996 *Phys. Rev. E* **53**(1) 926
[4] Tsai S-H, Wang F and Landau D P 2007 *Phys. Rev.* E **75** 061108
[5] Tsai S-H, Landau D P 2009 *Comp. Phys. Comm.* **180** 485
[6] Tsai S-H, Hu S, and Landau D P 2013 *Journal of Physics: Conference Series* current volume
[7] Freire, R T S, Mitchell, S J, Plascak J A and Landau, D P 2005 *Phys. Rev. E* **72** 056117
[8] Freire, R T S, Plascak, J A and Costa, B V 2004 *Braz. J. Phys.* **34**(2A) 438
[9] Landau DP and Binder K 2000 *A Guide to Monte Carlo Simulations in Statistical Physics* (Cambridge: Cambridge University Press)
[10] Plascak, J A, Ferrenberg, A M and Landau, D P 2002 *Phys. Rev. E* **65** 066702
[11] Wolff, U 1989 *Phys. Rev Lett.* **62** 361
[12] Metropolis, N, Rosenbluth, M N, Teller, J 1953 *Chem. Phys.* **21** 1087
[13] Ala-Nissila, T, Granato,E, Kankaala, K, Kosterlitz, J M and Ying, S -C 1994 *Phys. Rev. B* **50**, 12692
[14] Wilding, N B and Bruce, A D 1992 *J. Phys.: Condens. Matter* **4** 3087
[15] Ferrenberg, A M 1991 *Computer Simulation Studies in Condensed Matter Physics III* (Heidelberg: Springer-Verlag)
[16] Ferrenberg, A M and Swendsen 1988 *Phys. Rev. Lett.* **61** 2635
[17] Newman, M E J and Barkema, G T 1999 *Monte Carlo Methods in Statistical Physics* (Oxford: Clarendon)

Journal of Physics: Conference Series **487** (2014) 012007 doi:10.1088/1742-6596/487/1/012007

Combined molecular dynamics-spin dynamics simulations of bcc iron

Dilina Perera[1], David P Landau[1], Don M Nicholson[2], G Malcolm Stocks[2], Markus Eisenbach[2], Junqi Yin[2] and Gregory Brown[3]

[1] Center for Simulational Physics, The University of Georgia, Athens, GA 30602, USA
[2] Oak Ridge National Laboratory, Oak Ridge, TN 37831, USA
[3] Florida State University, Tallahassee, FL 32306, USA

E-mail: dilinanp@physast.uga.edu

Abstract.
Using a classical model that treats translational and spin degrees of freedom on an equal footing, we study phonon-magnon interactions in BCC iron with combined molecular and spin dynamics methods. The atomic interactions are modeled via an empirical many-body potential while spin dependent interactions are established through a Hamiltonian of the Heisenberg form with a distance dependent magnetic exchange interaction obtained from first principles electronic structure calculations. The temporal evolution of translational and spin degrees of freedom was determined by numerically solving the coupled equations of motion, using an algorithm based on the second order Suzuki-Trotter decomposition of the exponential operators. By calculating Fourier transforms of space- and time-displaced correlation functions, we demonstrate that the the presence of lattice vibrations leads to noticeable softening and damping of spin wave modes. As a result of the interplay between lattice and spin subsystems, we also observe additional longitudinal spin wave excitations, with frequencies which coincide with that of the longitudinal lattice vibrations.

1. Introduction

Computer simulations have played a vital role in advancing our knowledge and understanding of magnetic properties of materials. Investigation of model spin systems with Monte Carlo (MC) methods has revealed a substantial amount of information regarding the static thermodynamic properties of magnetic systems such as phase transitions and critical phenomena [1]. However, as the progression of "time" in these stochastic methods is completely unrelated to the "real time" associated with the true dynamics of the system, they are incapable of probing the deterministic time-dependent properties of the system. A widely used approach for exploring the dynamical behavior of magnetic systems is to numerically solve the classical equations governing the dynamics of magnetic moments at atomic level. This method, generally known as spin dynamics (SD) simulations, has been successfully employed in the past few decades to investigate the time-dependent properties of spin models with continuous degrees of freedom [2]. By choosing exchange interaction parameters determined from experiments or first principle calculations, one can construct realistic models of magnetic materials, where the obtained results can be quantitatively compared with the data from neutron scattering experiments [3, 4].

Recent, rapid advancements in SD simulations were largely due to the emergence of a novel class of symplectic integrators based on the Suzuki-Trotter (ST) decomposition of the exponential time evolution operator [5, 6]. These algorithms are known to be time reversible, and conserve phase-space volume exactly. In comparison with predictor-corrector schemes, decomposition methods are superior in terms of numerical stability, accuracy as well as efficiency [7].

Much of the theoretical and computational investigations of magnetic materials, by far, has been restricted to lattice-based spin systems, where the locations of the spins are fixed on regular lattice points. A major disadvantage of such simplified models is their lack of capacity to capture the wide range of complex phenomena that occur in real materials due to the coexistence of lattice vibrations (phonons) and spin waves (magnons). For instance, in transition metals, magnitudes of atomic magnetic moments and the strength of their interactions strongly depend on the changing local atomic environment in a non-trivial manner [8, 9]. On the other hand, magnetic interactions themselves play a significant role in determining the structural properties of such systems [10, 11]. Thus, the strong mutual interdependence between the lattice and spin subsystems suggest that any atomistic model designed to provide an accurate description of a real magnetic material should, in principle, facilitate the coupling between these subsystems.

To the best of our knowledge, the first attempt to incorporate the dynamics of translational degrees of freedom into a spin system was made by Omelyan *et al.* [12], in the context of a simple model for Heisenberg ferrofluids. By taking the exchange interaction as a pairwise function of inter-atomic distance, they presented a dynamical simulation model that could be exploited via combined molecular and spin dynamics simulations. For integrating the coupled equations of motion, Omelyan *et al.* devised a new algorithm based on the ST decomposition of the full time evolution operator which includes both translational and spin degrees of freedom. Using empirical many-body potentials and an exchange interaction parameterized by first principles calculations, Ma *et al.* [13] further extended the idea of coupled molecular and spin dynamics into a more realistic model for the treatment of ferromagnetic iron. They further investigated the rate of transfer of energy between lattice and spin subsystems via large-scale microcanonical (NVE) simulations and emphasized the shorter characteristic time scale associated with the dynamics of the spins in comparison to translational degrees of freedom.

Recently, Yin *et al.* [9] discussed the inadequacy of modeling the exchange interaction for itinerant systems as a simple function of inter-atomic distance since such models fail to describe the longitudinal fluctuations in magnetic moments. By introducing local atomic environmental parameters such as shell number and atomic volume, they developed a many-body-like exchange function for BCC iron based on first principles calculations. However, the proposed exchange function is not suitable for molecular dynamics simulations due to the complications associated with the discontinuity of its derivative at the cut-off distance.

In this paper, we present a compressible magnetic model for BCC iron, which is readily applicable for both Monte Carlo and combined molecular-spin dynamics simulations. We also discuss the generalization of the Suzuki-Trotter deposition method used for solving the coupled equations of motion for translational and spin degrees of freedom. Finally, by computing space-time Fourier transforms of correlation functions, we present solid evidence that indicate direct coupling between phonons and magnons.

2. Model and methods

2.1. Hamiltonian and equations of motion

Let us consider a system of N magnetic atoms of mass m, described by the classical Hamiltonian

$$\mathcal{H} = \sum_{i=1}^{N} \frac{m{\mathrm{v}_i}^2}{2} + U(\{\mathbf{r}_i\}) - \sum_{i<j} J(r_{ij})\mathbf{e}_i \cdot \mathbf{e}_j, \tag{1}$$

where $\{\mathbf{r}_i\}$ and $\{\mathbf{v}_i\}$ are positions and velocities respectively and $\{\mathbf{e}_i\}$ are the unit vectors representing the directions of atomic spins $\{\mathbf{s}_i\}$. $U(\{\mathbf{r}_i\})$ is the many-body atomic interaction term described by the Finnis-Sinclair potential for iron [14, 15]. The magnitudes of the spins $\mathbf{s}_i$ and $\mathbf{s}_j$ are absorbed into the exchange interaction $J(r_{ij})$, for which we take the same functional form and parameterization as used by Ma *et al.* [13].

Classical Hamilton equations yield the familiar equations of motion for positions and velocities as

$$\begin{cases} \dfrac{d\mathbf{r}_i}{dt} = \mathbf{v}_i \\ \dfrac{d\mathbf{v}_i}{dt} = \dfrac{\mathbf{f}_i}{m} \end{cases}, \tag{2}$$

where $\mathbf{f}_i = -\nabla_{\mathbf{r}_i} U_i + \sum_{j(j\neq i)} dJ(r_{ij})/dr_{ij}(\mathbf{e}_i \cdot \mathbf{e}_j)$ is the force acting the ith atom. Using the Poisson brackets method for the spin-dependent term of the Hamiltonian, one can obtain an equation for the precessional motion of the spins as,

$$\frac{d\mathbf{e}_i}{dt} = \left(\frac{g\mu_B}{\hbar M_i}\right) \mathbf{H}_i^{eff} \times \mathbf{e}_i \tag{3}$$

Here, $\mathbf{H}_i^{eff} = -\sum_{j(j\neq i)} J(\mathbf{r}_{ij})\mathbf{e}_j$ is the effective field on the ith spin, g is the electron g-factor, and $\{M_i\}$ are the magnitudes of atomic magnetic moments. A detailed discussion of the derivation of this formula can be found in Ref. [13].

Although the magnitudes of the moments are not considered as explicit phase variables in this model, they are allowed to vary dynamically via the dependence between the moment and the local atomic environment. Thus, whenever the atomic configuration changes during the simulation, moments $\{M_i\}$ are re-evaluated from a regression relationship between the moment and the local Voronoi volume of the atom, established via Locally Self Consistent Multiple Scattering (LSMS) first principles method [16].

2.2. Time integration algorithm

According to the Liouville formalism of classical mechanics, equations of motion can be represented as

$$\frac{dy(t)}{dt} = \hat{L}y(t), \tag{4}$$

where $y(t) = \{\mathbf{r}_i(t), \mathbf{v}_i(t), \mathbf{e}_i(t)\}$ denotes the complete set of phase variables consisting of positions, velocities and directions of the spins at time t. The Liouville operator $\hat{L}$ takes the form,

$$\hat{L} = \sum_{i=1}^{N} \left(\mathbf{v}_i \cdot \frac{\partial}{\partial \mathbf{r}_i} + \frac{\mathbf{f}_i}{m_i} \cdot \frac{\partial}{\partial \mathbf{v}_i} + \left(\frac{g\mu_B}{\hbar M_i}\right) \left(\mathbf{H}_i^{eff} \times \mathbf{e}_i\right) \cdot \frac{\partial}{\partial \mathbf{e}_i} \right). \tag{5}$$

The formal solution to Eq. 4 is

$$y(t+\tau) = e^{\hat{L}\tau} y(t), \tag{6}$$

where τ is the time step, and $e^{\hat{L}\tau}$ is the exponential operator which fully determines the time evolution of the microscopic phase variables. Unfortunately this operator cannot be evaluated exactly, thus one has to rely on approximate solutions which involve the decomposition of the operator into a series of sub-operators, where the respective operations are trivial enough to be evaluated analytically. As the first step, we write the Liouville operator $\hat{L}$ as a sum of three terms,

$$\hat{L} = \hat{L}_r + \hat{L}_v + \hat{L}_s, \tag{7}$$

where $\hat{L}_r = \sum_i \mathbf{v}_i \cdot \partial/\partial \mathbf{r}_i$ which only acts on positions, $\hat{L}_v = \sum_i \mathbf{f}_i/m_i \cdot \partial/\partial \mathbf{v}_i$ which only acts on velocities, and $\hat{L}_s = \sum_i (g\mu_B/\hbar M_i)(\mathbf{H}_i^{eff} \times \mathbf{e}_i) \cdot \partial/\partial \mathbf{e}_i$ which only operates on spin orientations. The solution given by Eq. 6 now takes the form,

$$y(t+\tau) = e^{(\hat{L}_r + \hat{L}_v + \hat{L}_s)\tau} y(t). \tag{8}$$

Grouping the operators $\hat{L}_r$ and $\hat{L}_s$ together as a single operator, one can use the second order Suzuki-Trotter formula [5] to decompose the exponential evolution operator as

$$e^{(\hat{L}_r + \hat{L}_v + \hat{L}_s)\tau} = e^{(\hat{L}_v + (\hat{L}_r + \hat{L}_s))\tau} = e^{\hat{L}_v \tau/2} e^{(\hat{L}_r + \hat{L}_s)\tau} e^{\hat{L}_v \tau/2} + \mathcal{O}(\tau^3),$$

which can be further decomposed by factorizing $e^{(\hat{L}_r + \hat{L}_s)\tau}$ in a similar manner to yield

$$e^{(\hat{L}_r + \hat{L}_v + \hat{L}_s)\tau} = e^{\hat{L}_v \tau/2} e^{\hat{L}_r \tau/2} e^{\hat{L}_s \tau} e^{\hat{L}_r \tau/2} e^{\hat{L}_v \tau/2} + \mathcal{O}(\tau^3). \tag{9}$$

Operators $e^{\hat{L}_r \tau}$ and $e^{\hat{L}_v \tau}$ merely shift the positions and and velocities of the atoms as [6]

$$\begin{cases} e^{\hat{L}_r \tau} y = \{\mathbf{r}_i + \mathbf{v}_i \tau,\ \mathbf{v}_i,\ \mathbf{e}_i\} \\ e^{\hat{L}_v \tau} y = \left\{\mathbf{r}_i,\ \mathbf{v}_i + \dfrac{\mathbf{f}_i}{m},\ \mathbf{e}_i\right\} \end{cases}. \tag{10}$$

Since the the effective field $\mathbf{H}_i^{eff}$ acting on each spin depends on all other spins that fall within its interaction range, a similar analytical form doesn't exist for the operator $e^{\hat{L}_s \tau}$. Therefore, one has to follow a ST decomposition scheme to find an approximate solution for $e^{\hat{L}_s \tau}$.

Krech *et al.* [7] and Tsai *et al.* [6] discuss the decomposition of the spin evolution operator $e^{\hat{L}_s \tau}$ for lattice-based spin systems by separating the lattice into a set of non-interacting sublattices. However, such decompositions cannot be directly applied to off-lattice spin systems where the neighbors of the spins are subjected to change with time. Therefore, following the method proposed by Omelyan *et al.* [12], we write the operator $e^{\hat{L}_s \tau}$ in terms of individual spin evolution operators as $e^{\hat{L}_s \tau} = \exp\left(\tau \sum_i \hat{L}_{s_i}\right)$ where $\hat{L}_{s_i} = (g\mu_B/\hbar M_i)(\mathbf{H}_j^{eff} \times \mathbf{e}_i) \cdot \partial/\partial \mathbf{e}_i$, and apply the second order ST decomposition iteratively to obtain

$$e^{\hat{L}_s \tau} = e^{\hat{L}_{s_1} \tau/2} \ldots e^{\hat{L}_{s_{N-1}} \tau/2} e^{\hat{L}_{s_N} \tau} e^{\hat{L}_{s_{N-1}} \tau/2} \ldots e^{\hat{L}_{s_1} \tau/2} + \mathcal{O}(\tau^3), \tag{11}$$

where the operator $e^{\hat{L}_{s_i} \tau}$ rotates the spin $\mathbf{s}_i$ by an angle $|\mathbf{H}_i^{eff}|\tau$ around $\mathbf{H}_i^{eff}$ while keeping all other spins fixed. Thus the operation $e^{\hat{L}_s \tau} y$ is performed as a series of spin rotations, strictly following the order specified by Eq. (11), which thereby leads to the exact conservation of spin lengths [7, 6].

2.3. Dynamic structure factor

In MD and SD simulations, time-dependent correlation functions of microscopic dynamical variables and their Fourier transforms are integral to the study of collective excitations, such as lattice vibrations and spin waves [17, 2]. Results of the simulations are usually conveyed as the dynamic structure factor $S(\mathbf{q}, \omega)$ for momentum transfer $\mathbf{q}$ and frequency transfer ω, a quantity that can be directly measured in inelastic neutron scattering experiments [18, 2].

Spatial Fourier transform of space-displaced, time-displaced density-density correlation function, namely the intermediate scattering function, is given by [19]

$$F_{nn}(\mathbf{q}, t) = \frac{1}{N} \langle \rho_n(\mathbf{q}, t) \rho_n(-\mathbf{q}, 0) \rangle, \tag{12}$$

where the angle brackets $\langle \cdots \rangle$ denote the ensemble average, and $\rho_n(\mathbf{q},t) = \sum_i \exp\left(-i\mathbf{q}\cdot\mathbf{r}_i(t)\right)$ is the spatial Fourier transform of microscopic atom density, $\rho_n(\mathbf{r},t) = \sum_i \delta\left(\mathbf{r}-\mathbf{r}_i(t)\right)$.

Similarly, by defining microscopic spin density as $\boldsymbol{\rho}_s(\mathbf{r},t) = \sum_i \mathbf{e}_i(t)\delta\left(\mathbf{r}-\mathbf{r}_i(t)\right)$, we write spin-spin intermediate scattering function as

$$F^k_{ss}(\mathbf{q},t) = \frac{1}{N}\left\langle \rho^k_s(\mathbf{q},t)\rho^k_s(-\mathbf{q},0)\right\rangle, \tag{13}$$

where $k = x, y,$ or z, and $\boldsymbol{\rho}_s(\mathbf{q},t) = \sum_i \mathbf{e}_i(t)\exp\left(-i\mathbf{q}\cdot\mathbf{r}_i(t)\right)$. By choosing a Cartesian coordinate system in the spin space such that the z axis is parallel to the magnetization vector, we regroup $F^k_{ss}(\mathbf{q},t)$ components considering the symmetry to obtain a longitudinal component

$$F^L_{ss}(\mathbf{q},t) = F^z_{ss}(\mathbf{q},t), \tag{14}$$

and a transverse component

$$F^T_{ss}(\mathbf{q},t) = \frac{1}{2}\left(F^x_{ss}(\mathbf{q},t) + F^y_{ss}(\mathbf{q},t)\right). \tag{15}$$

Fourier transforms of $F_{nn}(\mathbf{q},t)$, $F^L_{ss}(\mathbf{q},t)$ and $F^T_{ss}(\mathbf{q},t)$ yield the dynamic structure factor

$$S(\mathbf{q},\omega) = \frac{1}{2\pi}\int_{-\infty}^{+\infty} F(\mathbf{q},t)e^{-i\omega t}dt. \tag{16}$$

3. Results

3.1. Performance of the integration algorithm

Stability of any numerical integration scheme depends on the accuracy within which the conservation laws of the dynamics are observed over a period of time. The equations of motion for our model impose conservation of total energy, and for the special case of constant magnitudes for magnetic moments, conservation of magnetization as well. To illustrate the performance of our algorithm with regard to conservation laws, we have carried out microcanonical molecular-spin dynamics simulations on a $10 \times 10 \times 10$ simulation cell with periodic boundary conditions, starting from well equilibrated initial configurations at temperature $T = 800\,\mathrm{K}$.

At this point, we would like to elaborate more on the procedure followed in generating initial states for the time integration. Although our Hamiltonian directly allows transfer of energy between the lattice and spin subsystems through the coordinate dependence of the exchange interaction, due to the absence of a spin-orbit coupling term it does not facilitate transfer of angular momentum between the two subsystems. That is, in the absence of any external torques on the spins, the total spin angular momentum will remain constant, irrespective of any external perturbations applied to the lattice subsystem. As a consequence, thermostat algorithms used in conventional molecular dynamics simulations would fail to thermalize the spin subsystem to the desired temperature in the same way that they would equilibrate the lattice subsystem. Therefore, in order to prepare initial configurations for any finite temperature T, we adhere to the following procedure. As the first step, we equilibrate the subspace consisting of positions and spins using Metropolis Monte Carlo method [20]. Then we assign initial velocities to the atoms based on the Maxwell-Boltzmann distribution at the desired temperature. Finally, we carry out a microcanonical molecular-spin dynamics equilibration run which would ultimately bring the system to the equilibrium as a whole by resolving any inconsistencies between the position-spin subspace and the velocity distribution.

Fig. 1 shows energy [subset (a)] and magnetization [subset (b)] per atom as a function of time for different integration time steps. For the larger time steps considered ($\tau = 5\,\mathrm{fs}$ and $\tau = 10\,\mathrm{fs}$), we observe systematic drifts in the energy whereas with $\tau = 1\,\mathrm{fs}$, we were able

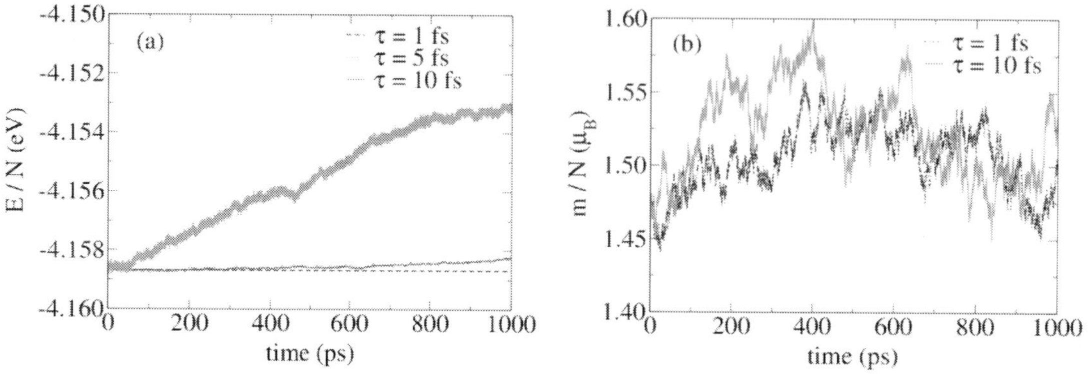

Figure 1. Time dependence of total energy (a) and magnetization (b) per atom for microcanonical molecular-spin dynamics simulations performed at temperature $T = 800\,\mathrm{K}$. The curves are for different integration time steps and the same initial configuration.

maintain the total energy conserved within an accuracy of six significant digits, without any noticeable drifts. Due to the isotropic nature of the spin-dependent term of our Hamiltonian, in principle, magnetization should also be a constant of motion. However, this conservation law breaks down in our model as a result of the volume dependence of the magnitudes of moments. Large fluctuations in the net magnetization shown in Fig. 1 for $\tau = 1\,\mathrm{fs}$ are merely consequences of this artifact, and should not be misinterpreted as numerical errors.

3.2. Effect of lattice vibrations on spin waves

To characterize collective excitations in our model, we have calculated density-density and spin-spin dynamic structure factors for $(q, 0, 0)$ direction at equilibrium temperatures $T = 293\,\mathrm{K}$ and $T = 800\,\mathrm{K}$. For finite linear lattice dimension L, only $q = 2\pi n_q/(La)$ with $n_q = \pm 1, \pm 2, \ldots, \pm L$ are allowed, where a is the lattice constant. Equations of motion were integrated up to $t_{\max} = 1.0\,\mathrm{ns}$ using a time step of $\tau = 1\,\mathrm{fs}$. The ensemble averages in Eq. (12) and (13) were computed using the results from 200 independent simulations, each initiated from an equilibrium configuration drawn from the canonical ensemble, using the procedure explained in section 3.1.

In Fig. 2, we have shown the transverse component of the spin-spin dynamic structure factor at temperature $T = 293\,\mathrm{K}$ for $|\mathbf{q}| = 0.66\,\text{Å}^{-1}$ [subset (a)] and $|\mathbf{q}| = 1.1\,\text{Å}^{-1}$[subset (b)]. Here, the full circles represent data obtained from molecular-spin dynamics simulations, whereas the open squares are from SD simulations performed with atoms frozen at perfect lattice positions. In order to make a fair comparison, exchange interaction parameters for the rigid lattice SD simulations were determined from the same pairwise function as used in molecular-spin dynamics simulations. In the presence of lattice vibrations, we observe that the peak positions of the transverse spin-spin dynamic structure factor systematically shift to lower frequencies. In addition, especially for larger $\mathbf{q}$ values, we find significant decrease in the peak amplitudes as well as increase in the half-widths, a behavior which clearly suggests damping of the spin waves. These effects of softening and damping of spin wave modes are in good qualitative agreement with the theoretical predictions by Woods *et al.* [21].

Fig. 3 shows the longitudinal component of the spin-spin dynamic structure factor for $|\mathbf{q}| = 0.22\,\text{Å}^{-1}$ at $T = 800\,\mathrm{K}$, for molecular-spin dynamics simulations [subset (a)] and rigid lattice SD simulations [subset (b)]. We conjecture that the multiple peak structure observed in the longitudinal component is a result of the two-spin-wave annihilation or creation, an effect which becomes more pronounced for small lattice sizes [22]. Curves in subset (a) and

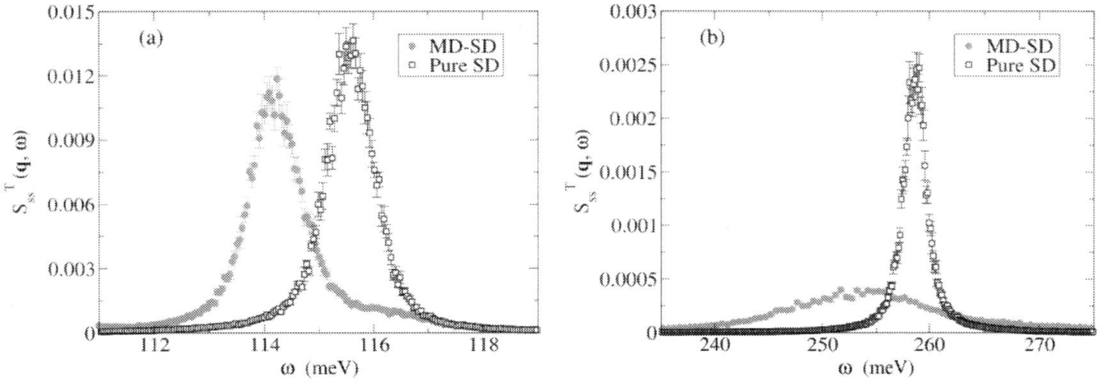

Figure 2. Transverse component of spin-spin dynamic structure factor $S_{ss}^{T}(\mathbf{q}, \omega)$ at temperature $T = 293\,\mathrm{K}$ for (a) $|\mathbf{q}| = 0.66\,\text{Å}^{-1}$ and (b) $|\mathbf{q}| = 1.1\,\text{Å}^{-1}$, along the $(q, 0, 0)$ direction for a $10 \times 10 \times 10$ simulation cell. Full circles (●) represent data obtained from molecular-spin dynamics (MD-SD) simulations while open squares (□) are from pure spin dynamics (SD) simulations performed on a rigid lattice.

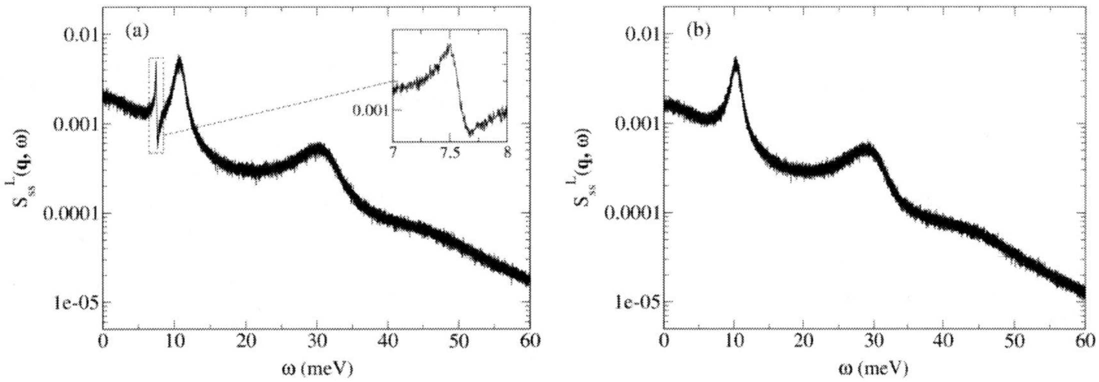

Figure 3. Longitudinal component of spin-spin dynamic structure factor $S_{ss}^{L}(\mathbf{q}, \omega)$ at $T = 800\,\mathrm{K}$ for $|\mathbf{q}| = 0.22\,\text{Å}^{-1}$, along the $(q, 0, 0)$ direction for a $10 \times 10 \times 10$ simulation cell. (a) Molecular-spin dynamics simulations (b) Pure spin dynamics simulations on a rigid lattice. The additional sharp peak at $\omega \approx 7.5\,\mathrm{meV}$ coincides with the peak position of the density-density structure factor for the same $|\mathbf{q}|$.

(b) are nearly identical, except for the sharp peak at $\omega \approx 7.5\,\mathrm{meV}$ which only appears in subset (a), for the case of molecular-spin dynamics simulations. The position of this peak coincides with the peak position of the density-density dynamic structure factor (Fig. 4) for the same $|\mathbf{q}|$ value, which suggest the existence of a propagating spin wave excitation coupled with the longitudinal phonon mode of the same frequency. We have observed similar coupled phonon-magnon excitations for all $|\mathbf{q}|$ values accessible through our simulations. Presence of longitudinal spin wave excitations at sound frequency is a new form of dynamical behavior in magnetic materials yet to be observed experimentally.

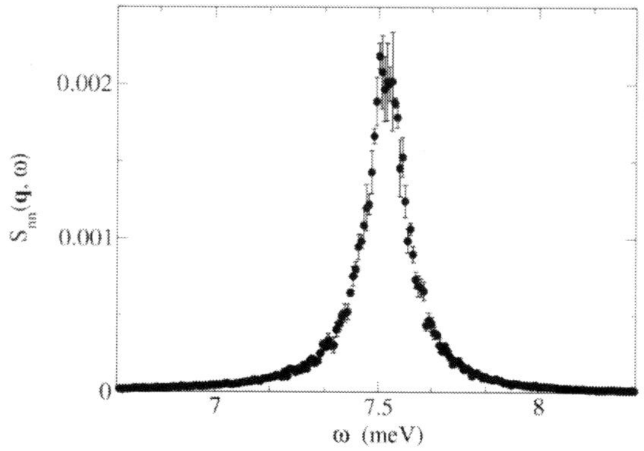

Figure 4. Density-density dynamic structure factor $S_{nn}(\mathbf{q},\omega)$ at $T = 800$ K for $|\mathbf{q}| = 0.22$ Å^{-1}, along the $(q, 0, 0)$ direction for a $10 \times 10 \times 10$ simulation cell.

4. Summary

In conclusion, we have used a compressible magnetic model to investigate the collective excitations in BCC iron. With the second order Suzuki-Trotter decomposition method we use for time integration, we were able to achieve stable numerical solutions over reasonably long time periods, with respect to the accuracy in energy conservation. In order to investigate the effect of lattice vibrations on spin wave excitations, we have computed spin-spin and density-density dynamic structure factors, and compared them to the results obtained from spin dynamics simulations performed on rigid lattices. In the presence of lattice vibrations, we observe softening and damping of spin wave modes. The additional peaks we observe in the longitudinal spin-spin dynamic structure factor coincide with the peak positions of the density-density dynamic structure factor, which indicates the possibility of longitudinal spin wave excitations at sound frequency.

Acknowledgments

Authors wish to thank Shan-Ho Tsai and Stefan Schnabel for informative discussions. This research was sponsored by the U.S. Department of Energy, Office of Basic Energy Sciences, Materials Sciences and Engineering Division, "Center for Defect Physics", an Energy Frontier Research Center. We also acknowledge the computational resources and technical expertise provided by the Georgia Advanced Computing Resource Center, a partnership between the University of Georgia's Office of the Vice President for Research and Office of the Vice President for Information Technology.

References

[1] Landau D P and Binder K 2009 *A Guide to Monte Carlo Simulations in Statistical Physics* 3rd ed (New York: Cambridge University Press)
[2] Landau D P and Krech M 1999 *J. Phys.: Condens. Matter* **11** R179
[3] Tsai S H, Bunker A and Landau D P 2000 *Phys. Rev. B* **61** 333
[4] Tao X, Landau D P, Schulthess T C and Stocks G M 2005 *Phys. Rev. Lett.* **95** 087207
[5] Suzuki M and Umeno K 1993 *Computer Simulation Studies in Condensed-Matter Physics VI* (*Springer Proceedings in Physics* vol 76) ed Landau D P, Mon K K and Schüttler H B (Springer Berlin Heidelberg) pp 74–86
[6] Tsai S H, Lee H K and Landau D P 2005 *Am. J. Phys.* **73** 615
[7] Krech M, Bunker A and Landau D P 1998 *Comput. Phys. Commun.* **111** 1–13
[8] Sabiryanov R F, Bose S K and Mryasov O N 1995 *Phys. Rev. B* **51** 8958
[9] Yin J, Eisenbach M, Nicholson D M and Rusanu A 2012 *Phys. Rev. B* **86** 214423
[10] Ekman M, Sadigh B, Einarsdotter K and Blaha P 1998 *Phys. Rev. B* **58** 5296
[11] Herper H C, Hoffmann E and Entel P 1999 *Phys. Rev. B* **60** 3839
[12] Omelyan I P, Mryglod I M and Folk R 2001 *Phys. Rev. Lett.* **86** 898

[13] Ma P W, Woo C H and Dudarev S L 2008 *Phys. Rev. B* **78** 024434
[14] Finnis M W and Sinclair J E 1984 *Phil. Mag. A* **50** 45
[15] Finnis M W and Sinclair J E 1986 *Phil. Mag. A* **53** 161
[16] Eisenbach M, Nicholson D M, Rusanu A and Brown G 2011 *J. Appl. Phys.* **109** 07E138
[17] Rapaport D C 2004 *The Art of Molecular Dynamics Simulation* 2nd ed (New York: Cambridge University Press)
[18] Lovesey S W 1984 *Theory of Neutron Scattering from Condensed Matter* (Oxford: Oxford University Press)
[19] Hansen J P and McDonald I R 2006 *Theory of Simple Liquids* 3rd ed (London: Academic Press)
[20] Metropolis N, Rosenbluth A W, Rosenbluth M N, Teller A H and Teller E 1953 *J. Chem. Phys.* **21** 1087
[21] Woods L M 2001 *Phys. Rev. B* **65** 014409
[22] Bunker A and Landau D P 2000 *Phys. Rev. Lett.* **85** 2601

Kosterlitz-Thouless Transition: The Diluted XY model

B.V. Costa
Department of Physics, Universidade Federal de Minas Gerais, Brazil
E-mail: bvc@fisica.ufmg.br

L.S. Lima
Department of Physics, Universidade Federal de Minas Gerais, Brazil

P. Z. Coura
Department of Physics, Universidade Federal de Juiz de Fora, Brazil

S. A. Leonel
Department of Physics, Universidade Federal de Juiz de Fora, Brazil

A.B. Lima
Department of Physics, Universidade Federal do Triangulo Mineiro, Brazil

Abstract. The effect of nonmagnetic disorder on Berezinskii-Kosterlitz-Thouless (BKT) transition in the two-dimensional ferromagnetic XY model on a square lattice, diluted by randomly placed vacancies can be an important source of information on the role played by vortexes in this transition. The transition temperature T_{BKT} is determined as a function of vacancy density ρ_v by stiffness ρ and for different size of the lattice. The anomalous behavior for T_{BKT} when ρ_v is close to the square lattice percolation limit is a consequence of the effect of finite size of lattice and not due the quantum fluctuations as concluded in literature. We obtain that T_{BKT} falls to zero at $\rho_v \approx 0.40$ close to the value 0.41 obtained for the classic model.

1. Introduction

The Beresinskii-Kosterlitz-Thouless theory [1, 2, 3] describes the physics of a very peculiar transition in the two-dimensional Coulomb gas and several related models like the super-fluid and superconductor films, the inter-facial roughening and the two-dimensional XY model among others [4, 5, 6]. In the BKT picture particles interact through a logarithm potential binding opposite charged particles. The transition is described as the unbinding of neutral pairs. In particular, in the two-dimensional XY model particles are vortex-anti-vortex pairs as seen in figure 1. Even being the BKT theory a well accepted phenomenological interpretation of the transition, Patrascioiu and Seiler gave a different perspective for the transition. They interpreted the transition as due to domain walls polymerization through the system [7]. With this interpretation they were capable of predicting the existence and the critical temperature of a phase transition in the Coulomb gas in any dimensionality $D > 1$. Numerical simulations corroborate their findings in two and three dimensions. So far, the phenomenology discussed by Patrascioiu and Seiler was put aside even if there is no conclusive arguments capable to decide in favor or not the *classical BKT* interpretation. We want to fix our attention in the BKT picture to build an argument that suggest the picture of unbinding of neutral pairs can not be responsible for the transition. To understand the nature of our argument we will spend a few lines to review the main features of the BKT picture using the XY model in two dimensions [8].

The XY model has a Hamiltonian given by

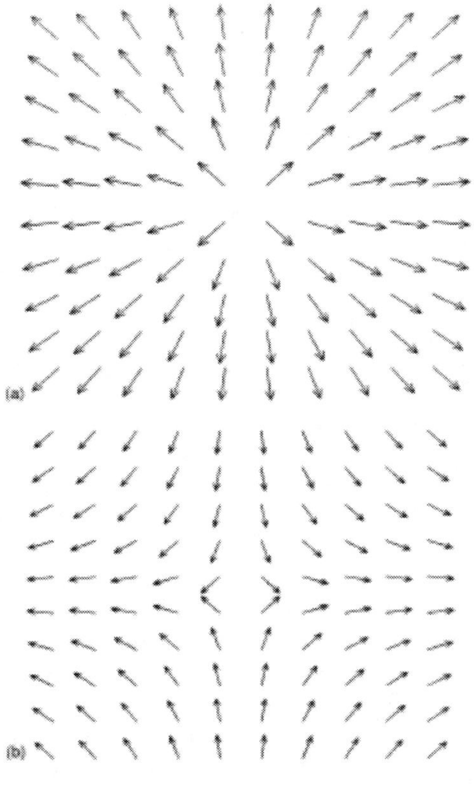

Figure 1. Schematic view of ferromagnetic vortexes configurations (top) and anti-vortexes (bottom) with winding numbers $q = +1$ and $q = -1$, respectively.

$$H_{XY} = J \sum_{<mn>} S^x_m S^x_n + S^y_m S^y_n, \tag{1}$$

with $\vec{S}_m = \{S^x_m, S^y_m, S^z_m\}$. Here, $< mn >$ indicates a sum over nearest-neighbor spins on a two dimensional square lattice. The Hamiltonian has an in plane symmetry. Instead of using this model in full we will put aside the out-of-plane S^z_m component of the spins for a while. This turns the Hamiltonian to

$$H_{PR} = J \sum_{<mn>} S^x_m S^x_n + S^y_m S^y_n = \sum_{<mn>} \cos\left(\theta_n - \theta_m\right), \tag{2}$$

the planar rotator. The correlation function [9] $e^{i\theta_0}e^{i\theta_{m-n}}$ can be estimated using a high temperature expansion as $e^{i\theta_0}e^{i\theta_{m-n}} \simeq exp\{- \mid m-n \mid ln(k_BT/J)\}$. On the other hand a low temperature expansion can be done by considering that the cosine argument does not fluctuates significantly. By retaining only terms up to second order one obtains in the large distance limit that $e^{i\theta_0}e^{i\theta_{m-n}} \simeq \mid n-m \mid^{-k_BT/2\pi J}$. Those arguments are in accordance with the Mermin-Wagner theorem which proves that the model never magnetizes. Beside that the low temperature expansion shows that the model has a critical line at low enough temperature characterized by a critical index $\eta = k_BT/2\pi J$. Those arguments, although not rigorous, suggest that the model has two phases without a true long range order. The nature of the transition was provided by Kosterlitz and Thouless [2]. They considered a Gaussian approximation for the Hamiltonian in equation 2 supplemented with periodicity in the Hamiltonian. The Hamiltonian leads to the equation of motion

$$\nabla^2\theta = 0, \qquad (0 < \theta < 2\pi) \tag{3}$$

Solutions to equation 3 are named vortexes (+) or anti-vortexes (−) and can be labeled by their winding numbers (See figure 1), q

$$\oint \nabla\theta \cdot d\ell = 2\pi q, \qquad q = 0, \pm 1, \pm 2, ... \tag{4}$$

An argument given by Kosterlitz and Thouless [2] suggest that vortices are irrelevant at low T but drive the transition at T_{BKT}. The argument starts by evaluating the energy of a configuration given by equations 3 and 4 using equation 2, $E = \pi J \ln(R/a_0)$, where R is the diameter of a vortex and a_0 is the lattice spacing. The energy diverges as $R \to \infty$. The entropy of a single vortex can be estimated as $k_B \ln(R/a_0)^2$, since a vortex is completly determined by giving the location of its origin. Therefore the free energy is obtained as $F = (\pi J - 2k_BT)\ln(R/a_0)$. The competition between the two terms shows that at $T = T_{BKT} = \pi J/2k_B$ the free energy is zero. At low temperature, the energy term gains the competition, so that, spin waves are the only relevant configurations. The spin-spin correlation function decays as a power law with distance. Free vortexes do not exist. On the other hand for all temperatures equal to or greater than T_{BKT} we may expect that the vortexes condensate in the ground state. The system is disordered and the correlation function falls off exponentially and free vortexes are expect to exist in the system. A more convincing argument is given by the renormalization group analysis provided by Kosterlitz and Thouless [2]. From this short review it is clear that to the BKT picture hold it is necessary that at (and above) T_{BKT} vortexes become free. If this is the case we could expect that a neutron scattering experiment should be able to observe their dynamics, since free vortex should disorder significantly the system. Several studies on neutron scattering on $stage-2CoCl2.Gic$ [10] have found strong indications of the expected BKT phase transition. They scanned the wave vector and frequency dependence of the spin correlation function in details. Above the T_{BKT} temperature they found the expected central peak in the in-plane correlation function. The out-of-plane correlation function showed only spin- wave peaks. Evertz and Landau [11] carried out a very extensive and careful spin dynamics simulation on the $2D-XY$ model. They found that the neutron-scattering function presented pronounced spin- wave peaks both in the in-plane and out-of-plane scattering functions over a wide range of temperatures. The in-plane scattering function also had a large number of clear but weak additional peaks which they interpreted as being from two-spin-wave process. In addition they observed a small central peak in the in-plane function at all temperatures, below and above T_{BKT}. In an attempt to describe the central peak, Mertens and co-workers [12] calculated the correlation function above T_{BKT}, assuming an ideal diluted gas of unbounded vortexes moving through the lattice. Such phenomenology was successful in describing the central peak in one-dimensional soliton dynamics in magnetic spin chains [13, 14, 15]. They found a squared Lorentzian central peak for the correlation function. However, the approach were not capable of explaining the experimental as well as the numerical results. Costa et al. using Monte Carlo and spin-dynamics simulation studied the vortex dynamics in the XY model. They calculated the lifetime of vortex-anti-vortex pairs, the time needed for a vortex to make a jump for one lattice spacing, the vortex density and the distance between pairs as a function of temperature as shown in figures 2 and 3. The results supported the idea that a vortex does not move through

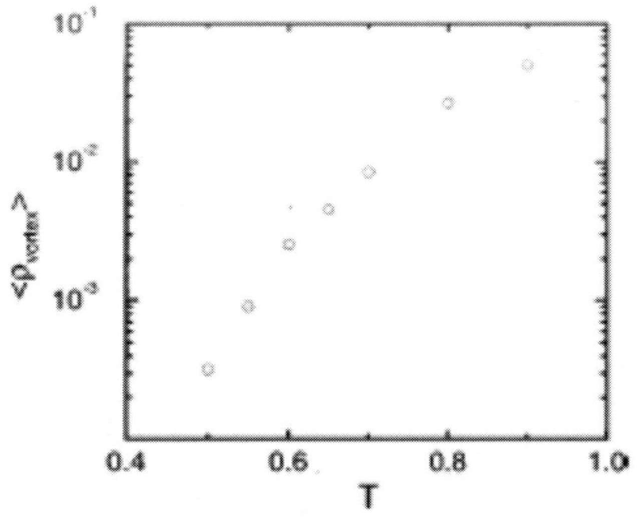

Figure 2. Vortex density as a function of temperature.

the lattice for long distances being a creation-annihilation process a more adequate picture to describe its " dynamics ". Of particular importance was the fact that their results did not found any " spectacular " change in the vortexes behavior when going through the BKT transition (See figure 3). The challenge then was to produce a reasonable phenomenology capable of explaining the central peaks observed in the simulations and experiments. More recently Costa and Lima [16] showed that the central peak in the two-dimensional XY model can be explained by a vortex-anti-vortex number fluctuation due to local diffusion and creation-annihilation processes. This behavior was seen more recently in the dynamics of vortexes in magnetic nano-dots [17]. An important lesson to learn from these fact is that the picture of free vortexes can no longer survive since it is not corroborated by experiments or simulations. However, the entropy can still be introduced by pairs fluctuation due to creation-annihilation processes [16]. To conclude our argument we have to go ahead a bit further. It is well known that vortexes become pinned in the presence of defects. For example, in diluted magnetic models when a site is empty, the vortex can be pinned there for a long period of time. It means that the creation-annihilation process will survive while any eventual dynamics will be severely suppressed in the presence of impurities before the percolation threshold concentration, p_c, is reached. It suggests that if the vortexes unbiding mechanism is responsible for the BKT transition we should expect the transition to extinguish at a temperature $T_{BKT}(p_c)$, with $p_c > p_{percolation}$. Here p is the concentration of magnetic sites. One can argue that the transition could change its character at p_c, however, the Harris argument [18] is valid here, so that, we may expect the transition will remain in the same universality class as the system is diluted up to the percolation threshold.

Journal of Physics: Conference Series **487** (2014) 012008 doi:10.1088/1742-6596/487/1/012008

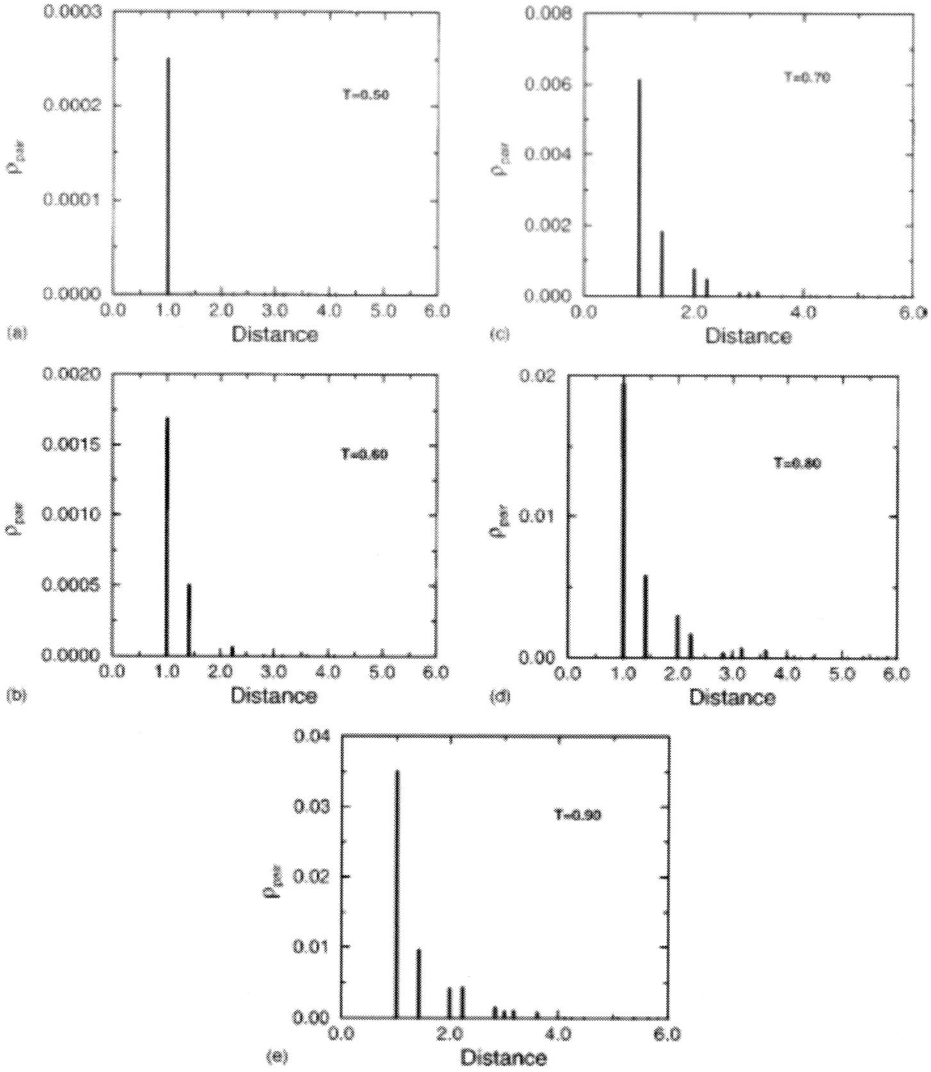

Figure 3. Number of the vortex-antivortex pairs as a function of the distance between them. Temperatures are indicated as inserts. The BKT transition is at $T_{BKT} = 0.700$

To go further we write the diluted XY model as

$$\mathcal{H} = J \sum_{\langle mn \rangle} \epsilon_m \epsilon_n (S_m^x S_n^x + S_m^y S_n^y), \tag{5}$$

with $J > 0$ and ϵ_m assumes the values 1 or 0 if the site is occupied by a magnetic or non-magnetic site respectively. The sum is over a $L \times L$ square lattice with periodic boundary conditions. In 1993 Lozovik and Pomirchi [19] reported some results for the Planar-Rotator model with bond dilution. They found that the BKT temperature behaves as $(\rho - \rho_{percolation})^{1.55}$, where ρ is the density of magnetic bonds. This result agreed very well with an earlier paper by D.C.

Harris et al. [20]. They found $(\rho - \rho_{percolation})^{\nu}$, with $\nu = 1.56024$. More recently Leonel et al. [21] obtained the phase diagram $T_{BKT} \times p$ for the site diluted model. Here, p is the density of magnetic sites. The critical percolation concentration is $p_c = 0.592746$. They found that at $p = 0.7$, far above the percolation threshold, the BKT was extinguished. Berche et al. [22] investigated the same model doing a more careful Monte Carlo simulation. They found that the apparent ill behavior obtained by Leonel et al. [21] was a fake, probably due to a poor statistics. This was confirmed by other result [23]. We did a much more careful numerical study [25] using lattice sizes $L \times L$, from $L = 16$ up to $L = 1280$, which is a much larger range than those used so far. The results indicated that the BKT transition remains up to the percolation threshold. We have used the stiffness to calculate the BKT temperature. The result is shown in figure 4.
More recently Young-Je et al. [24] measured the currentvoltage characteristics of site diluted

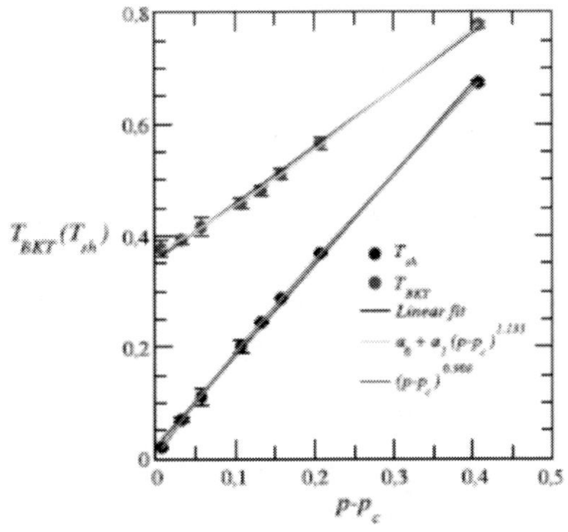

Figure 4. Plot of T_{BKT} and the maxima of the specific heat, T_{sh}, as a function of $p - p_c$. The points are the results of a finite size scaling for square lattices $(L \times L)$, with $L = 10, 20, 40, 80, 160, 320, 640$. Each point is obtained as an average over at least 10^6 configurations. Straight lines as well as power functions $(p - p_c)^q$ are adjusted to both curves. The power function adjusts much better to the simulated points than the straight lines. We have obtained $(p - p_c)^{1.133}$ and $(p - p_c)^{0.908} + c_0$ respectively.

Josephson junction arrays. They found evidences that the BKT transition is eliminated close to $p = 0.7$ and then a different type of order is established. This result is in flagrant disagreement with Refs. [22, 23] and the Harris argument [18]. The work done by Young-Je et al. [24] measuring the current-voltage in a site diluted Josephson junction arrays takes into account the quantum behavior of the system. We may argue that the differences found in contrast with the classical simulations may be due to quantum fluctuations as the BKT temperature is lowered close to zero. Having this in mind we have studied the spin transport of the diluted quantum XY model in two dimensions using the Self Consistent Harmonic Approximation, $SCHA$. We have used a Linear Spin Wave formalism described in [27] to study this model. The $SCHA$ consider quantum fluctuations as well as vortex fluctuations in the system. We have obtained that the spin conductivity falls abruptly to zero in the limit $p \approx 0.6$ changing the system from an ideal spin conductor to spin insulator (Figure 5). Changes in the behavior of the transition at $p = 0.7$ were not observed, indicating that the transition remains in the same universality class

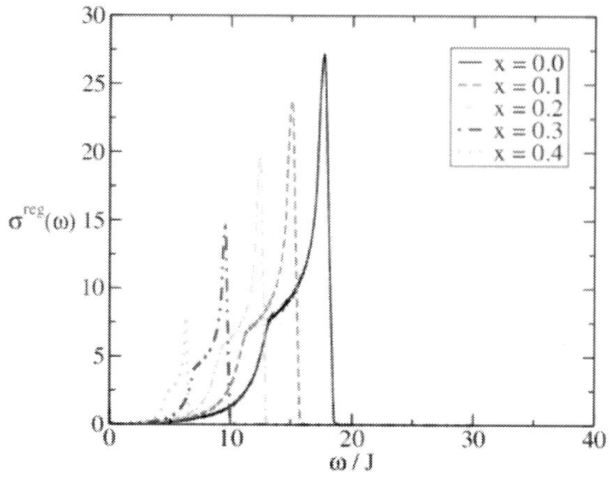

Figure 5. Spin conductivity of the diluted quantum XY model in two dimensions using the Self Consistent Harmonic Approximation, $SCHA$. We have used a Linear Spin Wave formalism described in [27] to study this model. The $SCHA$ consider quantum fluctuations as well as vortex fluctuations in the system. Here $x = 1 - p$. The spin conductivity does not show any qualitative change when going through $x = 0.3$.

from the beginning. To complete the study we have done a Quantum Monte Carlo study of the model using the ALPS code [26]. Similar results for the transition were found. Figure 6 show our findings. The arguments above indicate that the Kosterlitz-Thouless picture seems to not describe the entire phenomenology behind the BKT transition. Although our arguments are

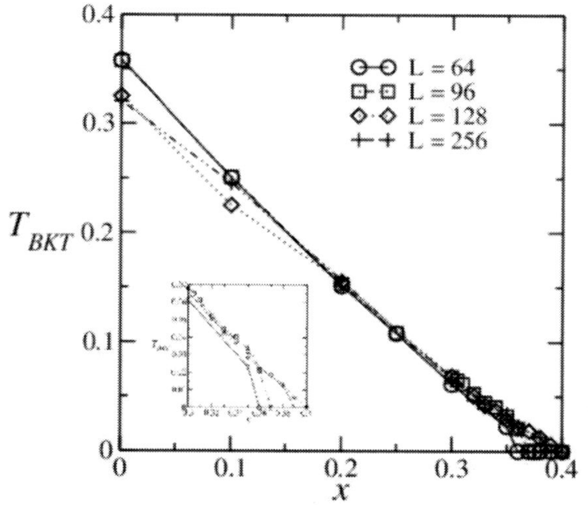

Figure 6. Quantum Monte Carlo results for the diluted quantum XY model in two dimensions using the ALPS code [26]. Here $x = 1 - p$. Although there is an abrupt drop of the BKT temperature before the critical concentration is reached, this occurs only for the smaller lattices, indicating that this is a finite size effect rather a true physical phenomenon.

quite reasonable, they need to be more convincing to be more palatable. We are now making a very careful simulation in the diluted XY model to study the behavior of vortexes as the dilution grows. We expect that as p approaches the percolation threshold the vortexes in the percolating cluster become pinned along the border of the cluster. If this is the case, those vortexes do not participate in the transition process and the BKT picture evidently fails. Having in sight the Harris argument, we are calculating the spin-spin correlation function inside the percolating cluster, so that, we can decide if the transition remains of the BKT type as expected.

Acknowledgment This work was partially supported by FAPEMIG and CNPq (Brazilian Agencies).

References

[1] V. Berezinskii, Zh. Eksp. Teor. Fiz. 59, 907 (1970) [Sov. Phys.-JETP 32, 493 (1971)].
[2] J.M. Kosterlitz and D.J. Thouless, Journal of Physics C 6, 1181(1973).
[3] Jorge V. José, 40 Years of Berezinskii-Kosterlitz-Thouless Theory (Singapore: World Scientific 2013).
[4] Petter Minnhagen, Reviews of Modern Physcs 59, 1001(1987).
[5] S.T. Chui and J.D. Weeks, Physical Review B 14, 4978(1976).
[6] M. Rapini, R. A. Dias and B. V. Costa, Physical Review B 75, DOI: 10.1103/PhysRevB.75.014425 (2007)014425.
[7] Adrian Patrascioiu and Erhard Seiler, Physical Review Letters 60, 875(1988).
[8] B. V. Costa, Brazilian Journal of Physics 41, DOI: 10.1007/s13538-011-0014-5 (2011)94-101.
[9] John B. Kogut, Reviews of Modern Physics 51, 659(1979).
[10] D.G. Wiesler, M. Suzuki, H. Zabel, S.M. Shapiro, R.M. Nicklow, Physica B and C 136 (1986) 22; D.G. Wiesler, H. Zabel, Physical Review B 36 (1987) 7303; D.G. Wiesler, H. Zabel, Journal of Applied Physics 63 (1988) 3554; D.G. Wiesler, H. Zabel, S.M. Shapiro, Physica B 156157 (1989) 292; H. Zabel, D.G. Wiesler, S.M. Shapiro, Zeitschrift fuer Physik B-Condensed Matter 93 (1994) 277.
[11] H.G. Evertz, D.P. Landau, Physical Review B 54 (1996) 12302.
[12] F.G. Mertens, A.R. Bishop, G.M. Wysin, C. Kawabata, Physical Review Letters 59 (1987) 117.
[13] J.A. Krumhansl, J.R. Schrieffer, Physical Review B 11 (1971) 3535.
[14] J.F. Currie, Physical Review A 16 (1977) 1692.
[15] H.J. Mikeska, Journal of Physics C 11 (1978) L29.
[16] B. V. Costa and A. B. Lima, Journal of Magnetism and Magnetic Materials 324, 1999-2005(2012).
[17] Colocar referencia de criacao-aniquilacao em nano-dot
[18] A. B. Harris, J. Phys. C 7, 1671 (1974)
[19] Yu.E. Lozovik, L.M. Pomirchi, Phys. Solid State 35 (1993) 1248.
[20] D.C. Harris, S.T. Herbert, D. Stroud, J.C. Garland, Phys. Rev. Lett. 67 (1991) 3606.
[21] S.A. Leonel, P.Z. Coura, A.R. Pereira, L.A.S. Mól, B.V. Costa, Phys. Rev. B 67 (2003) 104426.
[22] B. Berche, A.I. Farinas-Sanches, Y. Holovatch, R. Paredes, Eur. Phys. J. B 36 (2003) 91.
[23] G.M. Wysin, A.R. Pereira, I.A. Marques, S.A. Leonel, P.Z. Coura, Phys. Rev. B (2005) 094418.
[24] Young-Je, In-Cheol Baek, Mu-Yong Choi, Phys. Rev. Lett. 97 (2006) 215701.
[25] B.V. Costa, P. Z. Coura and S. A. Leonel, Physics Letters A 377 (2013) 1239-1241.
[26] A.F. Albuquerque, F.Alet, P.Corboz, P.Dayala, A.Feiguin, S.Fuchs, L.Gamper, E.Gull, S.Gurtler, A. Honecker, R. Igarashi, M. Korner, A. Kozhevnikov, A. Lauchli, S.R. Manmana, M. Matsumoto, I.P. McCulloch, F. Michel, R.M. Noack, G. Pawlowski, L. Pollet, T. Pruschke, U. Schollwock, S. Todo, S. Trebst, M. Troyer, P. Werner, S. Wessel, Journal of Magnetism and Magnetic Materials 310, 1187(2007).
[27] R. Schaffer, Anton A. Burkov, Roger G. Melko, Phys. Rev. B 80, 014503 (2009).

Short-time behavior of percolation observables at O(3) spin models: a preliminary analysis

Wanzeller W G
Universidade Federal da Fronteira Sul — Campus Laranjeiras do Sul, Paraná - Brazil
E-mail: wanderson@uffs.edu.br

Abstract. We analyzed the short-time behavior of percolation observables at three dimensional O(3) spin models. Following the Metropolis time evolution we can compare the magnetic and percolation order parameters. The magnetic order parameter evolves with a power law $M \sim t^{\theta}$, as expected. However the percolation order parameter follows a different law. We explored this particularity and raised some hypothesis. This paper is a preliminar analysis on this subject.

1. Introduction
In the literature, it is well-known that the magnetic order parameter (magnetization), at the critical point, evolves obeying a power law[1] $M \sim t^{\theta}$, where θ is a dynamic critical exponent and t is the (Monte Carlo) time. On the other hand, until now, the short-time evolution of percolation order parameter has received a few attention. This is due to, perhaps, these quantities present the same behavior at thermodynamic equilibrium[2] or because the theory of percolation does not have a proper dynamics.

Recently was shown that percolation order parameter and magnetization do not have the same behavior at the heat-bath short-time dynamics in the bidimensional Ising models[3]. This raised some questions about the evolution of the percolation order parameters to more complex spin models. In this work we will compare the short-time evolution of percolation and magnetic order parameters using the classical Heisenberg spin model [O(3)] and present some preliminary results about this question.

2. The classical Heisenberg spin models
The classical Heisenberg spin model is given by the hamiltonian (without external magnetic field)

$$\mathcal{H} = -\mathcal{J} \sum_{\langle ij \rangle} \vec{S}_i \cdot \vec{S}_j \,. \tag{1}$$

The constant $\mathcal{J}$ denotes the exchange interaction between nearest-neighbors $\langle ij \rangle$ three-dimensional unit spins $\vec{S}_i$ and $\vec{S}_j$. The magnetic order parameter is defined as the x-component of magnetization

$$M_x = \frac{1}{V} \sum_i S_i^x , \tag{2}$$

where V is the number of lattice sites (volume). This model presents a magnetic phase transition in a three-dimensional lattice at critical temperature[4],

$$T_c = 1.44292(8), \tag{3}$$

and many critical exponets[1, 5].

3. Percolation Observables

The mapping of a percolation problem in continuous spins models is well-known[9, 10] and will not be detailed here. In this work we study the short-time behavior of the percolation order parameter

$$\Omega = \frac{\Delta}{V}, \tag{4}$$

where Δ is the percolating cluster size (number of connected spins) and V the lattice volume.

4. Short-time dynamics

Using renormalization-group theory, it can be shown[6] that the early time evolution of an order parameter (e.g. the magnetization) already displays universal critical behavior, given by

$$M(t, \epsilon, m_0) = b^{-\beta/\nu}\mathcal{M}(tb^{-z}, \epsilon b^{1/\nu}, m_0 b^{x_0}), \tag{5}$$

where m_0 is the initial magnetization, $\epsilon \equiv (T - T_c)/T_c$, $\mathcal{M}$ is a universal function and b is a scale factor, which can be taken equal to $t^{1/z}$. We thus expect for $T = T_c$ and small m_0 a power-law behavior at early times

$$M(t)_{\epsilon\to 0} \sim m_0 t^{\theta}, \tag{6}$$

with $\theta = (x_0 - \beta/\nu)/z$.

5. Results

We have studied the three-dimensional case of the O(3) model, performing Monte Carlo simulations with 50,000 seeds (samples or histories) and 200 time steps for each seed. To analyse the short-time behavior of magnetic and percolation order parameters we simulated at $T = T_c$ for different values of the m_0 initial magnetization (0.002, 0.004 and 0.006). With the lattice initially hot (null magnetization) we randomically flipped some spins to obtain the magnetization m_0 (sharp). We can also prepare the initial lattice with $m_0 = 1$. Thus, the magnetization exhibits the power law $M(t) \sim t^{-\beta/z\nu}$. The same was done for the Baxter-Wu model [8]. Here this approach was not done yet.

In the implementation of O(3) algorithms we use the Metropolis one [7]. The x-magnetization [Eq. (2)] can be positive or negative. Thus, the percolating cluster can be positive or negative, so the percolation order parameter [Eq. (4)] is calculated from

$$\Omega = \Omega_+ - \Omega_-, \tag{7}$$

where Ω_+ (Ω_-) is the percolation order parameter for positive (negative) magnetization.

The data from Ω [Eq. (7)] are fitted by (see tables 1 and 2)

$$f(t)_\Omega = A\exp(-B/t), \tag{8}$$

where A and B are constants. This function is similar to the solution of the diffusion equation [3, 11]. On the other hand the magnetic order parameter follows a power law $M \sim t^{\theta}$ as expected. In figure 1 we have the time evolution of magnetic and percolation order parameters for $V = 60^3$ and $m_0 = 0.006$.

Table 1. Fit of percolation order parameter from Eq. (8) and $V = 30^3$.

$m_0 = 0.002$			$m_0 = 0.004$			$m_0 = 0.006$		
Δt	A	B	Δt	A	B	Δt	A	B
10-200	0.0020	19.9(1.4)	10-200	0.0046	22.3(0.7)	14-200	0.0076	24.5(6)
28-200	0.0022	26.0(2.9)	14-200	0.0048	24.1(0.8)	22-200	0.0075	27.1(7)
38-110	0.0624	32.9(6.1)	24-200	0.0050	26.3(1.0)	36-200	0.0070	29.2(1.2)

Table 2. Fit of percolation order parameter from Eq. (8) and $V = 60^3$.

$m_0 = 0.002$			$m_0 = 0.004$			$m_0 = 0.006$		
Δt	A	B	Δt	A	B	Δt	A	B
10-200	0.0028	118.1(3.6)	32-200	0.0057	118.4(1.9)	28-200	0.0086	119.1(1.7)
32-200	0.0030	124.4(3.6)	54-200	0.0060	124.1(2.8)	46-200	0.0091	126.7(1.9)
50-200	0.0032	134.4(4.8)				58-200	0.0094	131.3(2.4)

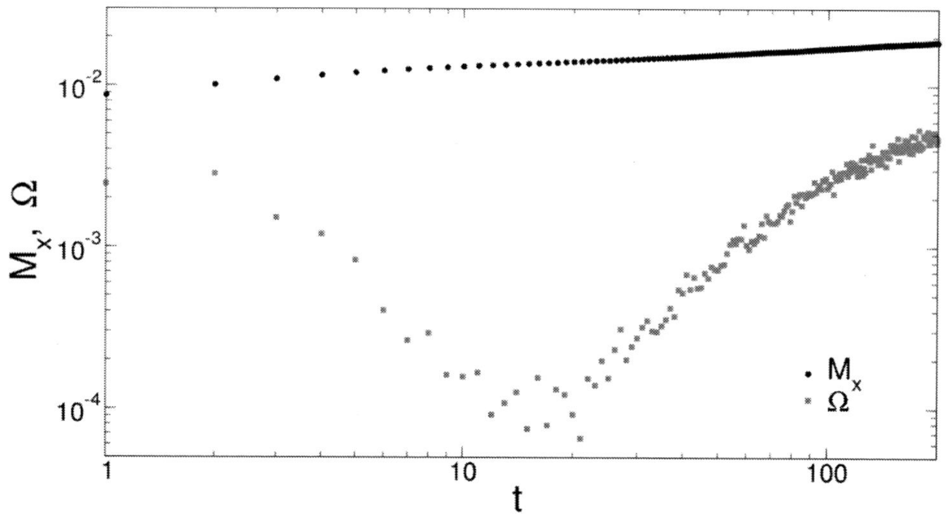

Figure 1. Comparison between time evolution of magnetic and percolation order parameters for $V = 60^3$ and $m_0 = 0.006$.

6. Conclusions

In this work we present a preliminary study about the short-time behavior of magnetic and percolation order parameters in the three dimensional Heisenberg model adopting Metropolis algorithm. From the data it is evident that the two observables do not exhibit the same behavior. Indeed, the magnetization follows a power law while the percolation order parameter follows a function solution similar to the diffusion equation. This fact has already been observed for the Ising model[3]. We must conclude our simulations soon using largest lattices and heat-bath algorithm[7, 12].

7. ACKNOWLEDGMENTS

The author thanks R. S. Marques de Carvalho for helpful comments and suggestions and Departamento de Informática em Saúde of Escola Paulista de Medicina - Universidade Federal de São Paulo, where part of the codes was executed.

References

[1] Zheng B 1998 *Int. J. Mod. Phys.* B 1419
[2] Wanzeller W G, Cucchieri A, Mendes T and Krein G 2004 *Braz. J. Phys.* **34** 247
[3] Wanzeller W G, Mendes T and Krein G 2006 *Phys. Rev.* E **74** 051123
[4] Chen K, Ferrenberg A M and Landau D P 1993 *Phys. Rev.* E **48** 3249
[5] Holm C and Janke W 1993 *Phys. Rev.* B **48** 936
[6] Janssen H K, Schaub B and Schmittmann B 1989 *Z. Phys.* B **73** 539
[7] Loison D, Qin C L, Schotte K D and Jin X F 2004 *Eur. Phys. J.* B **41** 395
[8] Santos M and Figueiredo W 2001 *Phys. Rev.* E **63** 042101
[9] Wolff U 1989 *Phys. Rev. Lett.* **62** 361
[10] Dimitrović I, Hasenfratz P, Nager J and Niedermayer F 1991 *Nucl. Phys.* B **350** 893
[11] Doremus R H 1995 *Rates of Phase Transformations* (New York: Academic Press).
[12] Miyatake Y, Yamamoto M, Kim J J, Toyonaga M and Nagai O 1996 *J.Phys.* C **19** 2539

Journal of Physics: Conference Series **487** (2014) 012010 doi:10.1088/1742-6596/487/1/012010

Parallel tempering and 3D spin glass models

T Papakonstantinou[1] **and A Malakis**[2]

[1,2] Department of Physics, Section of Solid State Physics, University of Athens, Panepistimiopolis, GR 15784 Zografos, Athens, Greece

E-mail: [1] tpapak@phys.uoa.gr; [2] amalakis@phys.uoa.gr

Abstract. We review parallel tempering schemes and examine their main ingredients for accuracy and efficiency. We discuss two selection methods of temperatures and some alternatives for the exchange of replicas, including all-pair exchange methods. We measure specific heat errors and round-trip efficiency using the two-dimensional (2D) Ising model, and also test the efficiency for the ground state production in 3D spin glass models. We find that the optimization of the GS problem is highly influenced by the choice of the temperature range of the PT process. Finally, we present numerical evidence concerning the universality aspects of an anisotropic case of the 3D spin-glass model.

1. Introduction

Parallel tempering (PT) or replica exchange is a Markov Chain Monte Carlo sampling method [1, 2, 3, 4, 5, 6, 7] that tends to become the method of choice in simulating complex systems where effective potentials have a complicated rugged landscape with many minima and maxima which become more pronounced with increasing system size [8, 9, 10, 11]. This behavior is encountered in spin glasses, where PT is widely used [12, 13, 14, 15, 16, 17, 18, 19, 20, 21, 22, 23, 24, 25, 26, 27, 28, 29, 30, 31, 32, 33, 34, 35]. Unlike Metropolis algorithm [36], where a single system is sampled in one temperature, in PT several replicas of the system evolve in their individual temperatures by local moves and exchange attempts between replicas are performed periodically (swap moves). This procedure gives PT the ability to roam more freely on the rugged energy landscape of complex systems. In PT, M replicas perform a number local moves, that is a number of MC steps not necessarily the same for each replica in their individual temperatures, and then an exchange (or swap) attempt between replicas is proposed. A successful implementation of PT sampling demands the selection of a suitable method for each part of the procedure and its fine-tuning. Initially one has to determine the temperature sequence of the replicas, then one has to decide which algorithm will be used for the local moves of the replicas, and how many local moves will be attempted for each replica before an exchange attempt. The algorithm for swapping replicas is also a basic ingredient of PT. Swapping algorithms between replicas which are adjacent in temperatures, is the most common practice. These algorithms will be denoted as $(NN)_x$, meaning nearest neighbor exchange according to some ordering specified by the sub-index x. Their alternatives are the all-pair exchange attempts (APE) methods and their recently proposed kinetic versions [37, 38].

In the present paper, we shall consider PT protocols with a variety of different ingredients. We test two methods for the temperature sequences. The widely used constant acceptance exchange (CAE) [6, 7, 39, 40], where the rate of successful swap attempts between replicas is constant, and the constant entropy increase (CEI) method introduced by Sabo *et al.* [41], where as its name declares requires a constant increase in entropy between successive temperatures. Three different algorithms are used to perform local moves. The Metropolis algorithm, the n-fold or BKL algorithm [42, 43, 44, 45], and the cluster algorithm of Wolff (W) [42, 46, 47]. Some APE and $(NN)_x$ algorithms for the replicas' swapping procedure are implemented and tested. Numerical tests, involving measurements for the specific heat errors and round-trip efficiency, are carried out for the square Ising model. Comparative tests are presented for the ground state problem in 3D spin-glass models. In our final section, we consider an anisotropic bimodal 3D spin glass model and discuss its universality properties. This model is a particular case of a more general (anisotropic) bimodal spin glass model [48, 49], defined by the Hamiltonian

$$H = -\sum_u \sum_{\langle ij \rangle_u} J^u_{ij} s_i s_j, \tag{1}$$

where the exchange interactions are uncorrelated quenched random variables, taking the values $\pm J^{xy}$ on the xy planes and the values $\pm J^z$ on the z axis. The bimodal distribution of J^u_{ij} takes the general form

$$P(J^u_{ij}) = p_u \delta(J^u_{ij} + J^u) + (1 - p_u)\delta(J^u_{ij} - J^u), \tag{2}$$

where u denotes the z axis ($u = z$) or the xy planes ($u = xy$), J^u denotes the corresponding exchange interaction strength and p_u are the probabilities of two neighboring spins (ij) having antiferromagnetic interaction. The standard isotropic 3D Edwards-Anderson bimodal (EAB) model [14, 12], corresponds to $J^z = J^{xy} = J(=1)$ and $p_z = p_{xy}$. The anisotropic case considered in our studies is the model $p_z = 0; p_{xy} \leq \frac{1}{2}$ with $J^z = J^{xy} = J(=1)$ [49].

2. Parallel Tempering (PT) Protocols

There is a rather large number of ideas that have been proposed in the last decade for the selection of temperatures in a PT protocol [37, 38, 40, 41, 50, 51, 52, 53, 54]. According to the approach followed by Katzgraber *et al.* [50], optimal temperature selection corresponds to the maximum rate of round trips between low and high temperatures in temperature space and can be obtained using a recursive readjustment of temperatures. Here we will consider two simpler but important methods. The CAE method, when used with appropriate number of sweeps between replica exchanges, was found to optimize the round-trip time [40]. To obtain the temperatures corresponding to a CAE rate r we follow here Ref. [40]. Starting from a chosen lowest temperature, adjacent temperatures are determined by calculating the acceptance exchange rate from

$$R(1 \leftrightarrow 2) = \sum_{E_1, E_2} P_{T_1}(E_1) P_{T_2}(E_2) p(E_1, T_1 \leftrightarrow E_2, T_2), \tag{3}$$

where $P_{T_i}(E_i)$ is the energy probability density function for replica i at temperature T_i and

$$p(E_1, T_1 \leftrightarrow E_2, T_2) = min[1, exp(\Delta\beta \Delta E)], \tag{4}$$

is the PT probability to accept a proposed exchange of two replicas, with $\Delta\beta = 1/T_2 - 1/T_1$ and $\Delta E = E_2 - E_1$. Demanding $R(1 \leftrightarrow 2) = r$ for all adjacent replicas, we obtain the temperatures of the required CAE sequence (from the above equations), provided that the energy probability density functions (PDFs) are known, or can be reasonably well approximated by some preliminary MC runs.

The second method requires a constant increase in entropy between successive temperatures [41]. Following Sabo *et al.* [41], we denote the M temperatures of the CEI sequence by $(T_m; m = 1, ..., M)$ and the total increase in entropy from T_1 to T_M by ΔS. Then the adjacent temperatures are determined starting from the given T_1 successively from

$$\int_{T_m}^{T_{m+1}} dT \frac{C_u(T)}{T} = \frac{\Delta S}{(M-1)}, \tag{5}$$

where the specific heat at any temperature can be calculated also from preliminary MC runs (single replica or PT) and histogram methods [46, 42].

Using such a preliminary PT run, we determine both CAE and CEI T-sequences in a temperature range centered around the pseudo-critical temperature of the specific heat of the square Ising model with linear size $L = 50$ ($N = L^D = L^2$ is the number of lattice sites). This is carried out, in a unified implementation, by appropriate recursive schemes, and then by repeating a Metropolis run at these T-sequences we estimate their canonical correlation times [42]. This practice can be applied to a general system for which the DOS is not known, applying in the first preliminary run (especially in a spin glass system) a PT protocol in an *ad-hoc* reasonable set of temperatures. For the CAE selection, we start from $T_1 = 1.9200$ (we set $J/k_B = 1$) and proceed to find higher temperatures corresponding to $r = 0.5$, reaching after a total number of temperatures $M = 19$ the temperature $T_M = 2.6903$. We also apply the CEI procedure with $M = 19$ between the same $T_1 = 1.9200$ and a final temperature $T_M = 2.6975$. Thus, the two schemes are defined approximately in the same temperature range with the same number of replicas.

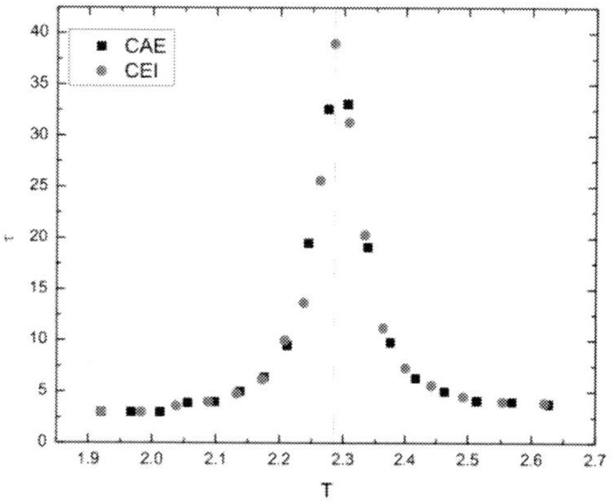

Figure 1. Temperature sequences for the CAE and CEI selection methods. Illustration of the corresponding canonical correlation times. The vertical dotted line indicates the specific heat maximum at $T_C^* = 2.285$ for $L = 50$.

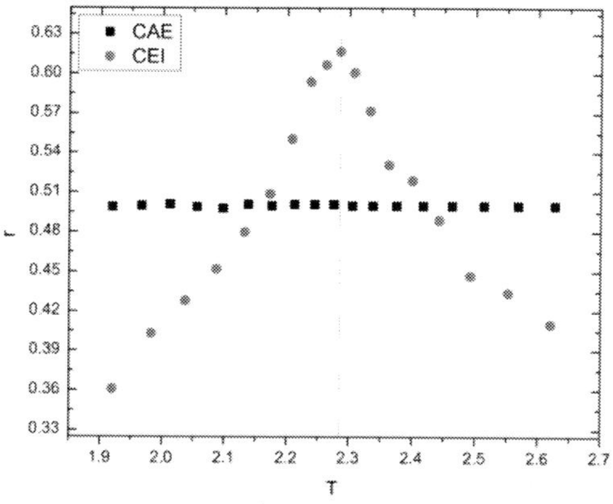

Figure 2. Illustration of the acceptance rates between adjacent replicas for the CAE and CEI selection methods. The vertical dotted line indicates again the specific heat maximum as in Figure 1.

Figure 1 illustrates the two T-sequences obtained according to the above description. In this illustration we display, for the CAE and CEI selection methods, the corresponding temperature sequences with their canonical correlation times, estimated from the discrete form of the energy autocorrelation function [42], in units of lattice sweeps (N Metropolis attempts). Figure 2 illustrates the acceptance rates between adjacent replicas for the two methods. The two methods produce temperature sequences that are more concentrated in the temperature range where the specific heat has a maximum (dotted lines in the figures), with the CEI method producing a more dense set close to the maximum point, and this is reflected in the variation of the acceptance rates of the CEI method.

We now discuss alternatives for the ordering of the exchange attempts. Mixing local MC attempts with swap attempts is essential in the PT procedure. It is this feature that enables an ergodic walk in temperature space, transferring information between the highest and lowest temperatures. Most PT protocols use NN exchanges, but recently PT protocols using APE methods have also been implemented [37, 38]. In a NN exchange protocol only adjacent exchange attempts are proposed. There are exactly $M-1$ different NN proposals and these may be uniquely denoted by the lowest temperature index $i = 1, ..., M-1$. We have implemented four different algorithms, denoted as $(NN)_a$,$(NN)_b$,$(NN)_c$, and $(NN)_d$, for the ordering of the NN proposals. In $(NN)_a$, a random permutation (say: $j_1, ..., j_{M-1}$) is generated, from the set $i = 1, ..., M-1$, and this permutation is used in an exchange swap cycle of $M-1$ proposals. Thus, swap moves are organized in cycles of $M-1$ proposals, and such a swap cycle (SC) may be used in defining the unit of time of an elementary PT step. In $(NN)_b$, the lowest temperature index (i) is randomly chosen from the set $i = 1, ..., M-1$ and therefore multiple exchanges of the same pair are allowed in the swap cycle. In $(NN)_c$, the sequence of proposals is fully deterministic, consisting of two swap sub-cycles, starting from the odd proposals $i = 1, 3, ...$, in increasing order, and continuing with the even proposals $i = 2, 4, ...$. This is a most widespread sequential update algorithm, supposed to speed-up the diffusion process. Finally, in $(NN)_d$, the odd and even sub-sequences are randomly permuted before starting the odd and following with the even swap sub-cycles.

In PT protocols using APE methods, adjacent or not adjacent replicas may be exchanged $(x_i \leftrightarrow x_j)$. In these methods, the number of possible proposals $(A \to B)$ is $M(M-1)/2$, $A = \{x_1, x_2, ..., x_i, ..., x_j, ..., x_M\}$ and $B = \{x_1, x_2, ..., x_j, ..., x_i, ..., x_M\}$. In the simplest case, for each proposal a pair (x_i, x_j) is randomly chosen from the set of all $M(M-1)/2$ different pairs. The generation of exchange attempts proceeds with equal probabilities, $P_{gen}(A \to B) = P_{gen}(B \to A) = 1/[M(M-1)/2]$, and the acceptance/rejection rule follows the Metropolis form, with an acceptance rate $P_{acc}(A \to B) = p(x_i \leftrightarrow x_j)$, given by Eq. (4). The swap attempts satisfy the detailed balance condition $P(A)P_{gen}(A \to B)P_{acc}(A \to B) = P(B)P_{gen}(B \to A)P_{acc}(B \to A)$, with $P(A) = \rho(x_1)\rho(x_2)...\rho(x_M)$, the product probability distribution, which becomes stationary with respect to the swap attempts. This is a simple APE method denoted here by APE_M, where the sub-index stands for the above Metropolis form of exchange acceptance rate. A different APE algorithm, denoted by APE_B, is the kinetic method of Brenner *et al.* [38]. The kinetic swapping procedure, involves explicitly only the $K = M(M-1)/2$ transitions of the form $A \to B \neq A$ $(j = 1, 2, ..., K)$ with selection probabilities $P_j = P_{acc}(A \to B)P_{gen}(A \to B)$, and $P_{gen}(A \to B) = 1/max\left\{\sum_{M\in\Phi} P_{acc}(A \to M), \sum_{L\in\Psi} P_{acc}(B \to L)\right\}$. Further details of such methods can be found in the original article [38] and in [55].

A PT step (PTS), is the elementary MC step used for the recording (measuring or averaging) process during an independent MC run. A PTS consists of one or several swap cycles of $M-1$ replica exchange attempts, and we shall be using the same definition, for simplicity, in the case of the APE method also. After each exchange attempt, all replicas perform a number of local moves (spin flips or cluster moves) at their current temperatures. The number of local moves $n(T_i)$, may vary with temperature and the total number of local moves (for any temperature in a SC) is $N_{local}(T) = (M-1)n(T)$. Without loss of generality, we will define the PTS to be just one SC. We may differentiate between various protocols by varying the number of local moves and we also are able to set $N_{local}(T) = f(T)\tau(T)N$, where $\tau(T)$ are the canonical correlation times and $f(T)$ are factors facilitating the adaption of the number of local moves $n(T)$. Note, that $N_{local}(T) = N$, corresponds to the standard lattice sweep. A first (disregarding) equilibration part of the simulation consist of t_{eq} PTSs, and then a large number of PTSs (t_{av}) is used for recording and averaging. Local MC moves are carried out by the Metropolis algorithm, the n-fold way algorithm and the cluster Wolff algorithm. We also use large numbers of independent MC runs (N_r) to obtain accurate estimates of our measures. In order to compare PT schemes, we also adapt time parameters in a consistent way so as to get approximately the same CPU time.

3. Comparing PT protocols

3.1. Testing PT protocols using the 2D Ising model

We now present some tests on PT protocols, obtained by combining a variety of features mentioned above. We begin with tests carried out on the L=50 square ferromagnetic Ising system with periodic boundary conditions. For this model we have used the exact DOS [56] to calculate the values of the specific heat at the PT protocol temperatures and thus we determine the following error-measures

$$\epsilon(T_i) = [C_{exact}(T_i) - C_{PT}(T_i)]/C_{exact}(T_i), \tag{6}$$

$$\bar{\epsilon} = \sum_{i=1}^{M} \epsilon(T_i)/M, \quad (7) \qquad \hat{\epsilon} = \sum_{i=1}^{M} |\epsilon(T_i)|/M, \quad (8) \qquad \epsilon^* = \max(|\epsilon(T_i)|), \quad (9)$$

For $(NN)_x$ exchange algorithms, which use the Metropolis algorithm as a local algorithm, we found that the above specific heat errors are very small and are all of the same order, with no statistically significant difference between them [55]. Thus, hereafter we shall use the $(NN)_a$ PT protocol and vary the other ingredients of the schemes.

Fig. 3, illustrates the errors $\epsilon(T_i)$ of two NN and two APE protocols using both the CEI and CAE selections. For all protocols in Fig. 3 the Metropolis algorithm has been used for the local moves. The number of independent MC runs is $N_r = 200$, with $t_{eq} = 3N$, $t_{av} = 15N$ and $N_{local}(T) = (M-1)n(T) = N$, where $M = 19$ is the number of temperatures. The error-measures, $\bar{\epsilon}$, $\hat{\epsilon}$, and ϵ^*, as defined above in Eq. (7), Eq. (8) and Eq. (9) are given, in parenthesis, on the panels of our figures. As can be seen from this figure, the specific heat errors of the $(NN)_a$ protocols are much smaller than the corresponding errors of the APE protocols. The CAE and CEI selections give comparable accuracy. The APE_B protocol suffers from very large specific heat errors that are more pronounced close to the specific heat maximum. This erratic behavior, is a subtle reflection of the kinetic character of the APE_B method [38]. On the other hand, the APE_M protocol, which is a standard PT protocol also allowing distant (Metropolis) exchange attempts, shows reasonable error behavior.

Let us now discuss PT cases using different local algorithms. Fig. 4 illustrates specific heat error behavior of PT protocols, performing local moves with the Wolff cluster algorithm (W), and the one-spin flip n-fold way or BKL algorithm. Only the CAE selection method has been used, and in an obvious notation, the protocols are denoted by $[BKL, (NN)_a]$, $[BKL, APE_B]$, $[W, (NN)_a]$, and $[W, APE_B]$, corresponding to $(NN)_a$ and the APE_B methods. The number of independent MC runs is $N_r = 200$. The protocols using the BKL algorithm, as local algorithm, use $t_{eq} = 3N$, $t_{av} = 15N$, and a reduced number of local moves $N_{local}(T) = (M-1)n(T) = 0.216N$. This choice corresponds to $n(T) = 30$ BKL spin flips for each replica after a swap move and makes the time requirement of the BKL protocols approximately equivalent to the corresponding Metropolis protocols. For the PT protocols using the Wolff algorithm, we used $t_{eq} = 3N/50$, $t_{av} = N$, and $N_{local}(T) = (M-1)n(T) = 0.0216N$, which corresponds to $n(T) = 3$ Wolff cluster flips after each exchange attempt. The adjustments are now reflecting the small dynamical exponent of the Wolff algorithm, and produce approximately the same time requirements as the PT protocols (of Fig. 3) using Metropolis algorithm.

By comparing the error-measures of the APE_B protocol of Fig. 3, which uses the Metropolis algorithm for the local moves, with the error-measures of the present APE_B protocols of

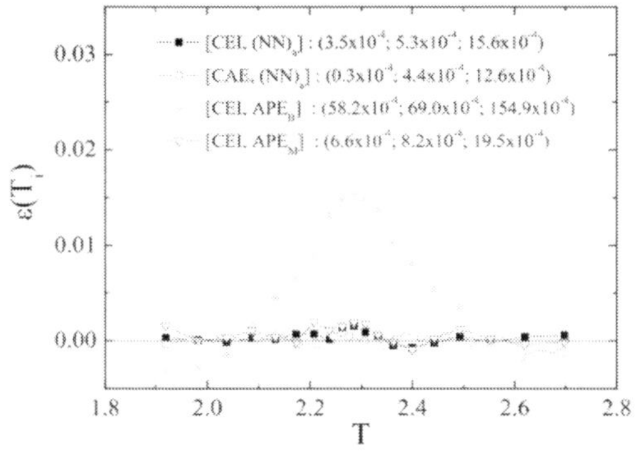

Figure 3. Specific heat errors for four different PT protocols using Metropolis algorithm for local moves. Error-measures, in parenthesis, as defined in Eq. (7), Eq. (8) and Eq. (9).

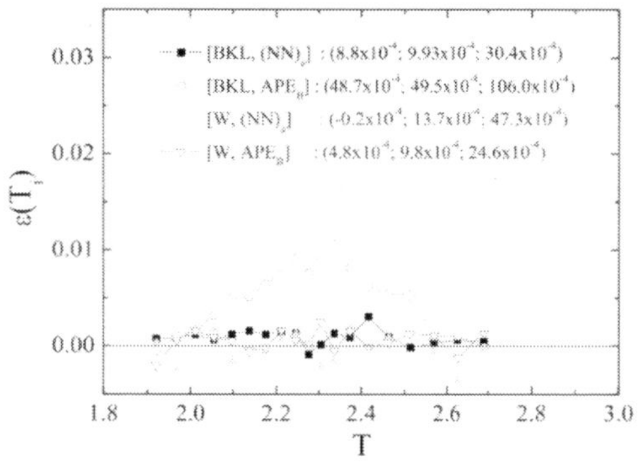

Figure 4. Behavior of PT protocols using n-fold way (BKL) algorithm and the Wolff cluster algorithm (W) for local moves. Only CAE selection of temperatures has been used.

Fig. 4, we observe clear improvement in the behavior of the illustrated error-measures. The improvement is substantial in the case of the $[W, APE_B]$ protocol and moderate in the case of the $[BKL, APE_B]$ protocol.

The efficiency of PT protocols is often characterized by measures indicating difficulties in the flow (bottleneck effects) as a replica moves from the high to low temperature and vise versa. Such a measure is the number of swap moves required to transfer any replica from the highest (lowest) temperature to the lowest (highest), and vise versa. Let the average number of exchange attempts required to transfer any replica from the highest (lowest) temperature to the lowest (highest) be denoted by u_j (d_j) for the corresponding transfer, averaged over the PTSs of a long independent run j ($j = 1, 2, ..., N_r$) and over the M different replicas of the PT protocol. Using these numbers we may define the average round-trip time $(u_j + d_j)$. Since, these quantities are strongly fluctuating (in the ensemble of N_r independent runs) it is more convenient to define and illustrate their behavior using running averages, defined bellow

$$[(u + d)]_k = \sum_{j=1}^{k} (u_j + d_j)/k, \tag{10}$$

where $k = 1, 2, ..., N_r$.

Figure 5 provides an illustration of the running averages of Eq. (10) for several PT protocols. All protocols in Figure 5 are using the Metropolis algorithm for local moves, $t_{eq} = 3N$, and $t_{av} = 15N$. The four protocols in the main panel use local time $N_{local}(T) = N$, while the three PT protocols in the inset use $N_{local}(T) = \tau(T)N$. In the notation, we have included also the numbers of independent runs $N_r = 200$ and $N_r = 20$ respectively as indicated in the legends (all protocols require approximately the same CPU time). By comparing the behavior in the panel and in the inset, we observe a strong influence of the number of local moves on the efficiency measure defined in Eq. (10). This is in accordance with the study of by Bittner *et al.* [40], verifying that a choice of the numbers of local moves, related to the canonical correlation times increases the efficiency of PT protocols. We also point out that the influence of temperature selection method is of minor importance, and that APE_B method yields a noticeable but not significant improvement.

Now, let us consider the behavior of the four PT protocols of Fig. 4, which use the Wolff cluster algorithm (W), and the BKL algorithm for local moves. Since, we have used different

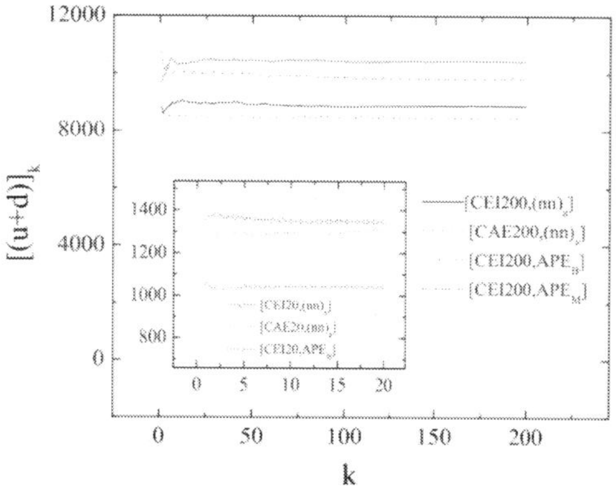

Figure 5. Running averages of round-trip time. Protocols in the main panel use $N_{local}(T) = N$, while in the inset $N_{local}(T) = \tau(T)N$.

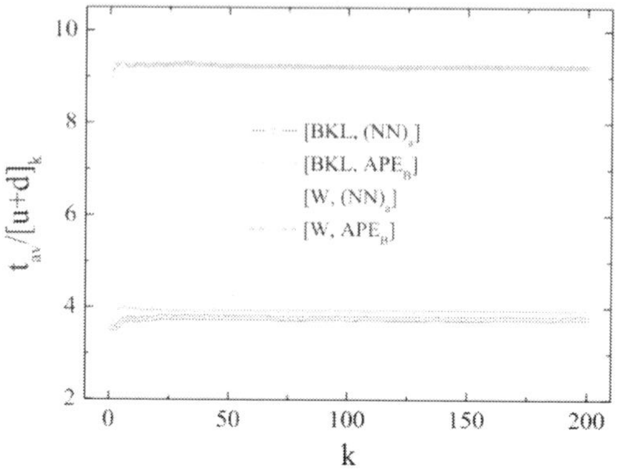

Figure 6. Running averages of $t_{av}/[u+d]_k$ ($\approx$ number of round-trips per replica).

t_{av}, we will now illustrate the behavior of running averages of the mean number of round trips per replica. The total number of swap moves in one independent run is $(M-1)t_{av}$, and thus $(M-1)t_{av}/(M[u+d]_k) \approx t_{av}/[u+d]_k$, measures approximately the mean number of round trips per replica. In Fig. 6, we have plotted the running averages of this quantity for the four PT protocols of Fig. 4, mentioned above. Comparing the two exchange methods we observe that, the APE_B method produces significant improvements in transfer efficiency. In the case of the $[W, APE_B]$ PT protocol, the improvement, compared with the $[W, (NN)_a]$ protocol, is very pronounced indicating also the strong influence of the local algorithm on the APE_B method. Of course, for each protocol the optimum in transfer efficiency will depend on a careful choice of the local times. The fast restoration of equilibrium that takes place after the Wolff local moves may be responsible for the improved behavior of the $[W, APE_B]$ protocol observed in both Fig. 4 and Fig. 6 . Thus, the tests on the Ising model have illustrated some of the merits and weaknesses of the selected PT schemes. However, the real power of PT methods, should be checked in rare-event problems in which the performance of conventional MC methods can become unreliable.

3.2. Testing PT protocols using the 3D EAB spin-glass model

Several studies have been carried out for determining true ground states (GS) of the (isotropic) 3D EAB model [11, 33, 34, 35, 57, 58]. Romá *et al.* [58] have addressed the question of whether it is more efficient, in order to find a true GS, to use large running times or several independent runs (t_{av} and N_r in our notation) for each realization of the disorder, called henceforth sample. The collapse example in Fig. 2b of their article, shows clearly that increasing the number of PTSs (t_{av}), has approximately the same effect with an analogous increase in the number of independent runs (N_r) for each sample. The production of a true GS depends essentially on the product $N_r t_{av}$. They have concentrated mainly on estimating the times necessary for generating true GS with a given probability. We have use this information in constructing the tests of this section, and also in undertaking the GS study for the anisotropic model in the next section.

We implement here two single spin-flip local algorithms, the Metropolis and the BKL algorithm, and try to observe the relative efficiency of the resulting protocols. We are interested in the dependence on the chosen local algorithm, and in the influence of the exchange method.

In order to illustrate the relative efficiency of the resulting PT protocols, we will first try to adjust the PT parameters in an approach requiring approximately the same CPU time for all protocols. This may be attempted by adjusting the local times of the protocols. However, as was pointed out in [55], the implementation of the APE_B method requires additional CPU time. This additional time comes from the involved recalculation of selection probabilities before each swap move. The relative time complexity of the method depends on the ratio between the local and averaging times, and increases considerably when we use a PT protocol with many temperatures. This obvious drawback of the method will be handled by restricting the possible remote exchanges only to significant exchanges (up to forth order), as was explained and implemented in [55]. Time differences, between $(NN)_a$ and APE_B exchange methods, remain considerable, even with the above restriction, in cases where the T-sequences consist of many temperatures. In such cases, we shall ignore the different time requirements of the exchange methods, and compare protocols with the same local time, because otherwise their adjustments may result in using inappropriately small local times for the APE_B method. We observe and compare their performances, albeit this practice favors the APE_B exchange method.

Each sample has its own peculiarities and these are reflected on the T-sequences obtained by a specified temperature selection method. In general, the dependence of T-sequence on the particular samples is very weak, and for our purposes it is more practical to use the same T-sequence for all samples. Thus, we average over T-sequences of a moderate number of $5-10$ samples, and then we use the averaged sequence for all samples included in our tests. Swap moves are attempted by $(NN)_a$, APE_M and APE_B methods. Numerical evidence for the performance of PT protocols, will be presented with the help of the sample averaged probability $[P]$ for producing true GS. This will be obtained from a relatively large set of N_s samples ($N_s = 100$), and from a relatively small set of N_s samples ($N_s = 5$). In the later case, as discussed in more detail bellow, the set is chosen to contain the hardest samples of the larger set. The sample averaged probability is defined by $[P] = \sum_{j=1}^{N_s} P_j/N_s$, where $P_j = n_j/N_r$ is the probability of reaching a GS for the jth sample. The probability P_j of reaching a true GS, is calculated from the n_j successful runs in a total number of independent N_r runs, carried out for each of the N_s samples.

Easy and hard samples exist with very different behavior [58], and in a large ensemble of samples the hardest samples have the smallest P_j. The sample averaged $[P]$, reflects the average performance of the PT protocols for the particular set of samples. If N_s is relatively large, the value of $[P]$ may be assumed to be a fair representative of efficiency measure for each protocol. However, we will be working in a regime of parameters, where for practically all relatively easy samples the probability P_j is ≈ 1.0. Note that, in order to verify that true GSs were found for each sample, we have used independent very long runs, with times longer than those derived by Ref. [58] for obtaining a true GS with probability ≥ 0.999. In the case of a small set of samples, $[P]$ is a less objective measure, but the choice of hardest samples magnifies the differences between the protocols, which are thus more easily observed.

In [55], we have presented comparative tests for the 3D EAB spin-glass, carried out for a cubic lattice with linear size $L = 6$. We found that the results for the CAE T-sequence were slightly better, but not significantly different, from the corresponding results for the CEI T-sequence. Furthermore, the superiority of the BKL algorithm, compared to the Metropolis algorithm, was firmly established. Also, we observed that the all-pair exchange APE_M method deteriorates the production process of GSs, while the APE_B method produced a noticeable improvement. Here, we shall use, in all our tests, a cubic lattice with linear size $L = 8$ and T-sequences only by the CAE selection method.

In our first test, we consider an ensemble of $N_s = 100$ samples and implement a temperature sequence corresponding to a constant acceptance rate $r = 0.1$, producing as explained above, an

Journal of Physics: Conference Series **487** (2014) 012010 doi:10.1088/1742-6596/487/1/012010

averaged sequence of $M = 9$ temperatures in the range $T = 0.4 - 2.5$. The difference between BKL and Metropolis updates is taken into account by adjusting the local time by a factor of the order ≈ 4, in favor of the corresponding Metropolis PT protocols. For the simplest protocol $[M,(NN)_a]$, we set $N_{local}(T) = (M-1)n(T) = N(=L^3)$, which corresponds to $n(T) = 64$ Metropolis spin-flip attempts for all replicas after each swap move. Local time for the other protocols is determined by short preliminary runs. In a particular run with $t_{av} = 5N$ and $N_r = 600$, we used $n(T) = 64$, $n(T) = 18$, and $n(T) = 16$ for the $[M,(NN)_a]$, $[BKL,(NN)_a]$, and $[BKL,APE_B]$ protocols resulting in 1742, 1675 and 1806 CPU minutes respectively. Thus, in our first test we tried to compare PT protocols using approximately the same CPU time by varying t_{av}.

The behavior of the protocols $[M,(NN)_a]$, $[M,APE_B]$, $[BKL,(NN)_a]$, $[BKL,APE_B]$, and $[BKL,APE_M]$, by varying the running time parameter t_{av}, is illustrated in Fig. 7. APE_B method, in the $[BKL,APE_B]$ protocol, produces a marginal improvement in the large t_{av} range, when compared with the corresponding NN protocol $[BKL,(NN)_a]$. On the other hand APE_M method, in the $[BKL,APE_M]$ protocol, seems to worsen the production process of GSs, when compared with the $[BKL,(NN)_a]$ protocol. The superiority of the BKL local algorithm is very clearly revealed by comparing the corresponding PT protocols with those using a Metropolis local algorithm. Noteworthy is also that n-fold way updates have been also the basis of some previous searches for GSs of spin glasses [59, 60, 61]. The picture emerging from the above test is very similar with that discussed in [55]. Nevertheless, APE_B method was found, in our study of the 2D Ising model, to be sensitive to the local time used, as can be seen also in Fig. 5. APE_B may also be sensitive to the spacing of temperatures, in the temperature sequence used, since it is possible that denser sequences may facilitate remote exchanges. We consider these possibilities in the following tests.

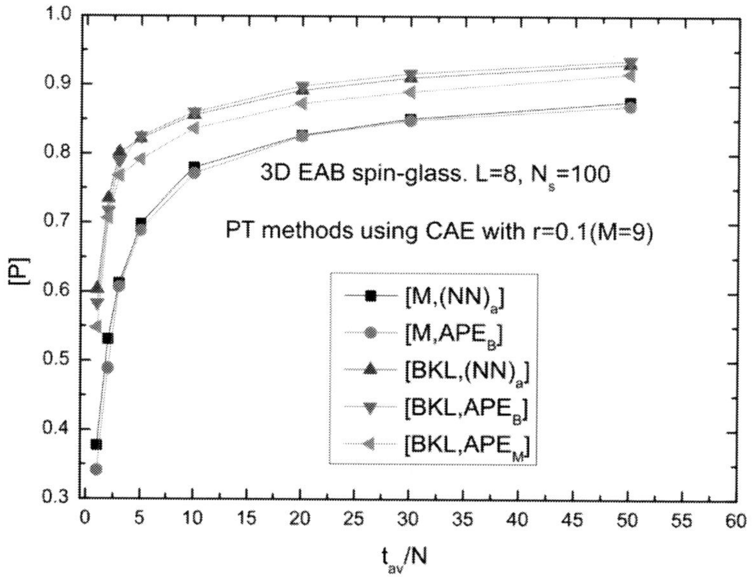

Figure 7. Performance of PT protocols, with local time adjusted to produce approximately the same CPU time.

We now ignore the additional CPU time of the APE_B method, and try to compare it with the $(NN)_a$ method, using exactly the same local time in the corresponding PT protocols. We consider the performance of only the two most efficient protocols $[BKL,(NN)_a]$, and

$[BKL, APE_B]$ protocols and also we implement a more dense T-sequence, the CAE T-sequence with $r = 0.5$, corresponding to $M = 21$ temperatures (approximately in the same temperature range). The dependence of the efficiency of APE_B method on local time is first inspected. We fix the number of PTSs to a value $t_{av} = 10N$, and vary the local time as follows: $N_{local}(T) \approx (0.24N), 5(0.24N), 10(0.24N), 20(0.24N)$. Since $M = 21$, the number of BKL spin-flips for all replicas after each swap attempt is correspondingly $n(T) = 6, 30, 61, 122$. In order to magnify the difference between the two exchange methods, we also restrict our tests only in the set of the five hardest realizations chosen from the original ensemble of $N_s = 100$ samples. Fig. 8, demonstrates that, as we increase local times APE_B method becomes more efficient from $(NN)_a$ method. This is, of course, in agreement with the behavior observed in Fig. 4. APE_B method appears more effective when the exchanging spin configurations are less correlated. However, the actual effectiveness of the APE_B for the GS-problem is still undisclosed, as we shall see bellow. From Fig. 8, APE_B method is already superior to $(NN)_a$ method when $N_{local}(T) = 5(0.24N)$ or $n(T) = 30$. We now fix local time in that value, and vary the running time: $t_{av} = 2N, 5N, 10N, 20N, 30N, 40N$. Fig. 9 shows the performance of the two exchange methods in this test, which we shall call test A. For the set of the five hardest realizations, APE_B method appears again to be more efficient in finding true GS, although we have to take into consideration the fact that the local time chosen favors the APE_B method.

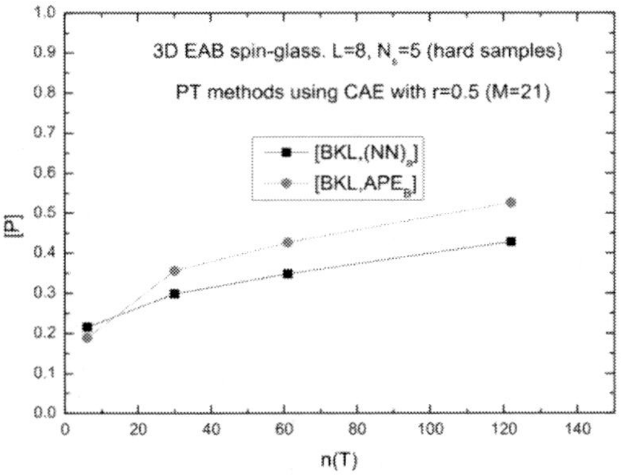

Figure 8. Performance of the exchange methods as we vary local time $n(T)$, with fixed $t_{av} = 10N$.

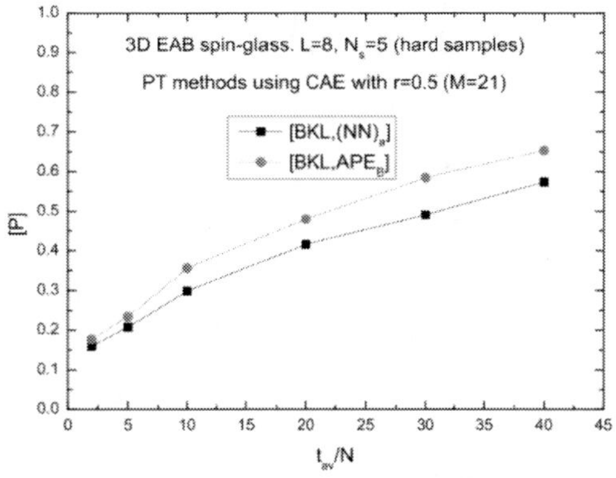

Figure 9. Performance of the exchange methods (test A), with fixed local time $n(T) = 30$.

In the next test, which we call test B, we use again the same CAE T-sequence ($M = 21$), but now we fix the local time to a lower level ($n(T) = 6$) and vary the averaging time: $t_{av} = 10N, 25N, 50N, 100N, 150N, 200N$. A further third test, test C, was undertaken, in which we consider only the lowest $M = 7$ temperatures of the $r = 0.5$ CAE T-sequence, in a T-range: $T = 0.4 - 0.967$. In test C, we use local time $n(T) = 20$ and vary the averaging time: $t_{av} = 10N, 25N, 50N, 100N, 150N, 200N$. Illustration in Fig. 10, contains the performance of the two exchange methods for the three tests in a new x-scale, which is simply the total number of BKL spin-flips divided by N, namely $M(M-1)n(T)t_{av}/N$. This is traditionally called total number of BKL Monte Carlo steps (MCS), in one independent run of the protocols. The new scale facilitates the comparison between points corresponding to the same value on the x-axis, since differences in CPU time are now due only to the additional CPU time of the APE_B method. Both exchange methods, $(NN)_a$ and APE_B, in tests B and C, produce comparable results for the probability $[P]$ of the selected five hardest samples, and therefore one may insist in using the simplest and faster $(NN)_a$ approach. Both tests B and C give higher values of $[P]$

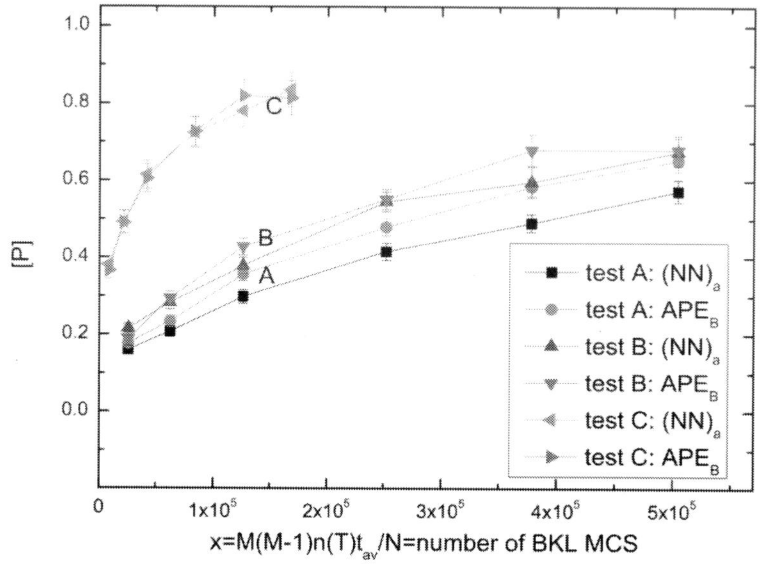

Figure 10. Performance of exchange methods for tests A, B and C. All protocols use BKL updates. Test C involves the lowest $M = 7$ temperatures of the $r = 0.5$ CAE T-sequence with $M = 21$ temperatures.

than those of test A, and the previous minor superiority of the APE_B method in Fig. 9 appears now as an insignificant incidence of the particular choice of local time. It is very interesting that C test, is by far the most efficient in producing true GS with much larger probabilities $[P]$.

To summarize, considering also the CPU time differences of the exchange methods, we observe that test A, in which the APE_B method appears to be a moderate winner, is not an optimum choice for the production of GS. In the other two tests, the $(NN)_a$ and APE_B exchange methods produce comparable results. The search for true GS depends on several parameters of the PT schemes, including the above mentioned ingredients, the number N_r of independent runs and, of course, the temperature range used. A comprehensive study and critical examination of all these factors appears necessary before optimality is claimed. Careful choice of the temperature range emerges now as one of the most influential features for the optimization of the GS-problem, which we believe that has not been recognized in previous studies.

4. Universality aspects of a 3D anisotropic spin-glass model

Recently, the spatially uniaxial anisotropic 3D EAB model $p_z = 0; p_{xy} \leq \frac{1}{2}$ with $J^z = J^{xy} = J(=1)$, has been studied and its phase diagram has been estimated [49]. The phase diagrams for both the isotropic Edwards-Anderson model and the present anisotropic model are reproduced here in Fig. 11. The vertical solid line in this figure, indicates a ferromagnetic-paramagnetic (F-P) phase transition for $(p_z = 0; p_{xy} = 0.176)$. For this case, a detailed finite size scaling analysis (FSS) was carried out in [49], and Fig. 12, illustrates the behavior of effective exponents for the correlation length exponent at this point. The relevant estimation gives $\nu = 0.683(3)$, which is in excellent agreement with the estimate $\nu = 0.6837(53)$ of the extensive numerical investigations of Ballesteros *et al.* [62] for the site-diluted Ising model. This and similar findings for the other critical exponents have demonstrated very clearly the universality for the ferromagnetic-paramagnetic transitions, between the standard isotropic EAB model and the present anisotropic model [49].

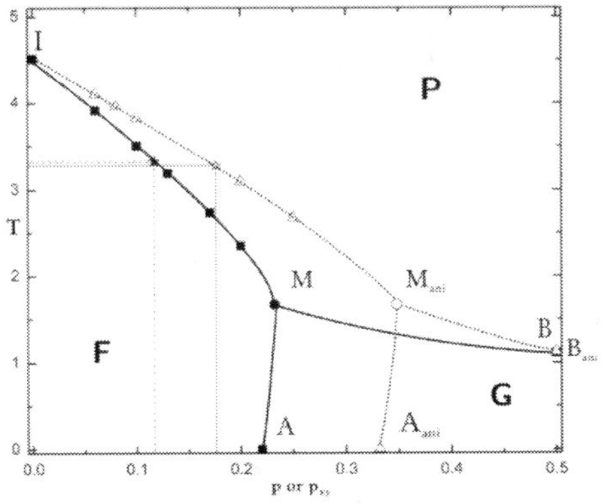

Figure 11. Phase diagrams for the isotropic Edwards-Anderson model (solid lines and full symbols) and the present anisotropic model $p_z = 0, p_{xy}$ (dash lines and open symbols) [49].

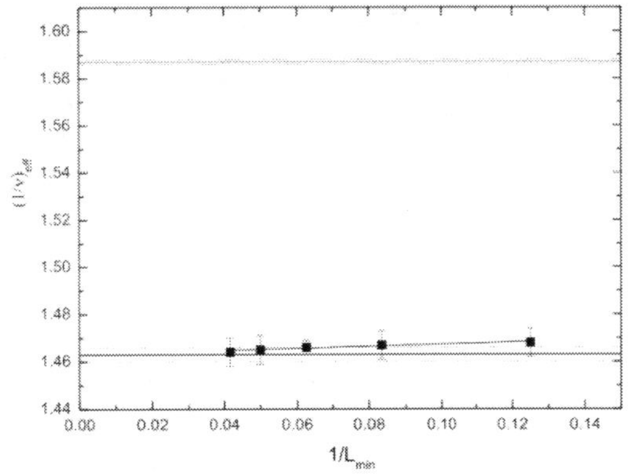

Figure 12. Effective exponents $(1/\nu)_{eff}$ for the anisotropic model, giving by linear extrapolation $1/\nu = 1.463(3)$ [49]. The estimation of the pure 3d Ising model [63] is also shown for comparison.

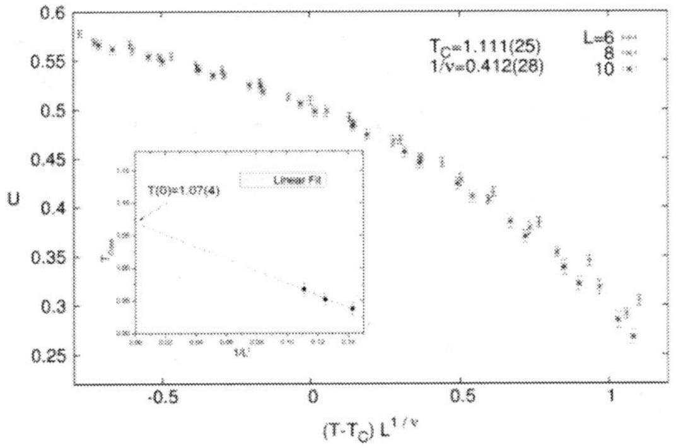

Figure 13. Two methods for the estimation of the phase diagram point B_{ani}. A linear extrapolation of crossings points of the Binder's fourth order cumulant, and a collapse attempt on these cumulants.

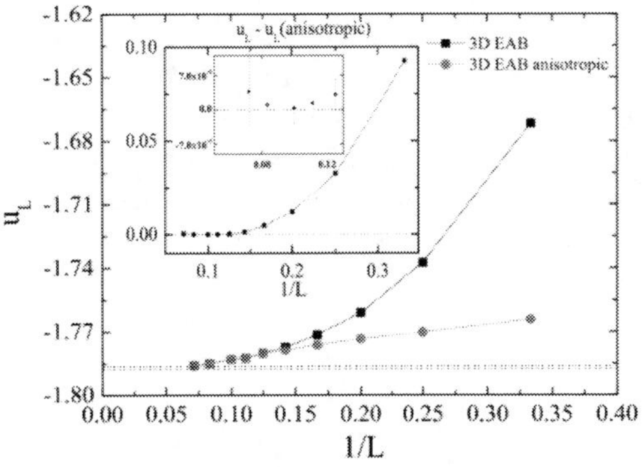

Figure 14. Finite-size behavior of GS energies per site for the 3D EAB model and the present anisotropic model. Their differences are smaller than the estimated errors.

The phase diagram point B_{ani} (open triangle on the phase diagram) corresponding to $(p_z = 0, p_{xy} = \frac{1}{2})$, appears to coincide with the isotopic phase diagram point B corresponding to $(p_z = p_{xy} = \frac{1}{2})$. Attempts to estimate the temperature of the B_{ani} point appear in Fig. 13. The numerical evidence presented is based on data for the Binder's fourth order cumulant ($U = 1 - \frac{1}{3}\left(\frac{[\langle q^4\rangle]}{[\langle q^2\rangle]^2}\right)$) where $\langle \ldots \rangle$ and $[\ldots]$ denote thermal and disorder average respectively and q the spin-glass overlap order parameter ($q = \frac{1}{N}\sum_{i=1}^{N} s_i^{\alpha} s_i^{\beta}$, where s_i denotes the spin of site i and "α" and "β" represent two replicas of the same disorder realization). The inset illustrates a linear extrapolation of the crossings points of Binder's fourth order cumulants and gives the estimate $T_{B_{ani}} = 1.07(4)$. In the main panel of Fig. 13, a data collapse for these cumulants provide the estimate $T_{B_{ani}} = 1.111(25)$, very close to the corresponding estimate of the isotropic case $T_B = 1.109(10)$ [26]. The critical exponent $1/\nu$ in the panel of Fig. 13 corresponds to the phase diagram point B_{ani}. This gives for the correlation length exponent an estimate $\nu = 2.43(15)$, in fair agreement for the exponent estimate obtained for the isotropic case (phase diagram point B) $\nu = 2.39(5)$ given by Katzgraber et al. [32].

Finally, we consider the problem of ground states for the anisotropic model at $(p_z = 0, p_{xy} = 0.5)$. Following the approach of [58], we produced estimates for the finite-size GS energy per site for the anisotropic model. Our simulations cover the range of sizes $L = 3 - 14$, using the PT protocol $[CAE, BKL, (NN)_a]$. In all runs, we choose $N_r = 1$, use a short disregarding part ($t_{eq} = 2N$) and vary the rest of the PT parameters in a way that the main running times (t_{av}) are comparable to those in Table B.3 of [58]. Fig. 14, presents the finite-size behavior of the GS energy per site for both isotropic and anisotropic 3D EAB models. In the insets, their differences are illustrated. In particular, it shown that their differences, for $L \geq 6$ are much smaller than the estimated errors. Therefore, the asymptotic limit of these GS energies practically coincide. The two dashed lines in the main panel indicate previous asymptotic estimations, namely $u_\infty = -1.7863(4)$ [64] and $-1.7876(3)$ [65].

5. Summary and Conclusions

We reviewed several PT schemes and examined their accuracy and efficiency. Our tests on the 2D Ising model suggest that the two methods of selecting temperature sequences (CAE and CEI), produce results that are accurate and are almost equivalent in efficiency. Efficiency of PT protocols is greatly increased by using number of local moves related to the canonical correlation time. Wolff algorithm, when used for the local moves, greatly increases the efficiency of the PT protocols. In particular, APE_B method of Brenner *et al.* [38], when used with local Wolff updates, was found both accurate and very efficient. Using the problem of GS production of the 3D EAB model, we carried out a further comparative study of PT protocols. Various implementations of the APE methods were compared with protocols based on the $(NN)_a$ method, and a critical discussion on the routes for improving performance was carried out. We also found that, a careful choice of the temperature range used for the PT is one of the most influential features for the optimization of the GS-production.

Finally, we presented numerical evidence concerning some universality aspects of an anisotropic case of the 3D spin-glass model. The emerging picture, from the illustrated universality for the ferromagnetic-paramagnetic and spin-glass-paramagnetic transitions, between the standard isotropic EAB model and the anisotropic model, supports the view that the universality between the two models is a general feature covering the global phase diagram of Fig. 11. The finite temperature phase diagram points between spin-glass and paramagnetic phases appear to be very close, and we also presented numerical evidence that the asymptotic limit of the GS energy of the isotropic ($p_z = p_{xy} = 0.5$) and the anisotropic ($p_z = 0, p_{xy} = 0.5$) EAB models, are also very close, and possibly coincide.

Acknowledgments

This work was supported by the special Account for Research of the University of Athens.

References

[1] Landau D P and Binder K 2000 *Monte Carlo Simulations in Statistical Physics* (Cambridge: Cambridge University Press)
[2] Swendsen R H and Wang J-S 1986 *Phys. Rev. Lett.* **57** 2607
[3] Hukushima K and Nemoto K 1996 *J. Phys. Soc. Jpn.* **65** 1604
[4] Marinari E, Parisi G and Ruiz-Lorenzo J J 1998 *Spin Glasses and Random Fields*, ed A P Young (Singapore: World Scientific) p 59
[5] Marinari E 1996 arXiv:cond-mat/9612010v1
[6] Earl D J and Deem M W 2005 *Phys. Chem. Chem. Phys* **7** 3910

[7] Katzgraber H G 2009 arXiv:0905.1629v3
[8] Fiore C E 2011 *J. Chem. Phys.* **135** 114107
[9] Fiore C E and da Luz M G E 2011 *Phys. Rev. Lett.* **107** 230601
[10] Fiore C E and da Luz M G E 2013 *J. Chem. Phys.* **138** 014105
[11] Alvarez Baños R et al. 2010 *J. Stat. Mech.* P06026
[12] Nishimori H 2001 *Statistical Physics of Spin Glasses and Information Processing: An Introduction* (New York: Oxford University Press)
[13] Binder K and Kob W 2005 *Glassy Materials and Disordered Solids* (Singapore: World Scientific)
[14] Edwards S F and Anderson P W 1975 *J. Phys.* F **5** 965
[15] Nishimori H 1980 *J. Phys.* C **13** 4071
[16] Nishimori H 1986 *J. Phys. Soc. Japan* **55** 3305
[17] Ozeki Y and Nishimori H 1987 *J. Phy. Soc. Japan* **56** 3265
[18] Le Doussal P and Harris A B 1989 *Phys. Rev.* B **40** 9249
[19] Singh R R P 1991 *Phys. Rev. Lett.* **67** 899
[20] Ozeki Y and Ito N 1998 *J. Phys.* A **31** 5451
[21] Palassini M and Caracciolo S 1999 *Phys. Rev. Lett.* **82** 5128
[22] Kawashima N and Rieger H 2003 arXiv:cond-mat/0312432v2
[23] Jörg T 2006 *Phys. Rev.* B **73** 224431
[24] Hasenbusch M, Parisen Toldin F, Pelissetto A and Vicari E 2007 *Phys. Rev.* B **76** 184202
[25] Campbell I A, Hukushima K and Takayama H 2007 *Phys. Rev.* B **76** 134421
[26] Hasenbusch M, Pelissetto A and Vicari E 2008 *Phys. Rev.* B **78** 214205
[27] Ceccarelli G, Pelissetto A and Vicari E 2011 *Phys. Rev.* B **84** 134202
[28] Billoire A, Fernandez L A, Maiorano A, Marinari E, Martín-Mayor V and Yllanes D 2011 *J. Stat. Mech.* P10019
[29] Hukushima K, Takayama H and Yoshiro H. 1998 *J. Phys. Soc. Jpn.* **67** 12
[30] Ballesteros H G, Cruz A, Fernandez L A, Martín-Mayor V, Pech J, Ruiz-Lorenzo J J, Tarancon A, Tellez P, Ullod C L and Ungil C 2000 *Phys. Rev.* B **62** 14237
[31] Katzgraber H G, Palassini M and Young A P 2001 *Phys. Rev.* B **63** 184422
[32] Katzgraber H G, Körner M and Young A P *Phys. Rev.* B **73** 224432
[33] Moreno J J, Katzgraber H G and Hartmann A K 2003 *International Journal of Modern Physics* C **14** 285-302
[34] Katzgraber H G and Young A P 2003 *Phys. Rev.* B **67** 134410
[35] Katzgraber H G, Körner M, Liers F, Jünger M and Hartmann A K 2005 *Phys. Rev.* B **72** 094421
[36] Metropolis N, Rosenbluth A W, Rosenbluth M N, Teller A H and Teller E 1953 *J. Chem. Phys.* **21** 1087
[37] Calvo F 2005 *J. Chem. Phys.* **123** 124106
[38] Brenner P, Sweet C R, VonHandorf D and Izaguirre J A 2007 *J. Chem. Phys.* **126** 074103
[39] Sugita Y and Okamoto Y 1999 *Chem. Phys. Lett.* **314** 141
[40] Bittner E, Nußbaumer A and Janke W 2008 *Phys. Rev. Lett.* **101** 130603
[41] Sabo D, Meuwly M, Freeman D L and Doll J D 2008 *J. Chem. Phys.* **128** 174109
[42] Newman M E J and Barkema G T 1999 *Monte Carlo Methods in Statistical Physics* (Oxford: Clarendon)
[43] Bortz A B, Kalos M H and Lebowitz J L 1975 *J. Comput. Phys.* **17** 10
[44] Schulz B J, Binder K and Muller M 2001 *Int. J. Mod. Phys.* C **13** 477
[45] Malakis A, Martinos S S, Hadjiagapiou I A and Peratzakis A S 2004 *Int. J. Mod. Phys.* C **15** 729
[46] Swendsen R H and Wang J-S 1987 *Phys. Rev. Lett.* **58** 86
[47] Wolff U 1989 *Phys. Rev. Lett.* **62** 361
[48] Güven C, Berker A N, Hinczewski M and Nishimori H 2008 *Phys. Rev.* E **77** 061110
[49] Papakonstantinou T and Malakis A 2013 *Phys. Rev.* E **87** 012132
[50] Katzgraber H G, Trebst S, Huse D A and Troyer M 2006 *J. Stat. Mech.: Theory Exp.* p03018
[51] Kofke D A 2002 *J. Chem. Phys.* **117** 6911; 2004 *J. Chem. Phys.* **120** 10852
[52] Kone A and Kofke D A 2005 *J. Chem. Phys.* **122** 206101
[53] Predescu C, Predescu M and Ciobanu C V 2004 *J. Chem. Phys.* **120** 4119
[54] Neirotti J P, Calvo F, Freeman D L and Doll J D 2000 *J. Chem. Phys.* **112** 10340
[55] Malakis A and Papakonstantinou T 2013 *Phys. Rev.* E **88** 013312
[56] Beale P D 1996 *Phys. Rev. Lett.* **76** 78
[57] Weigel M *Phys. Rev.* E **76** 066706
[58] Romá F, Risau-Gusman S, Ramìrez-Pastor A J, Nieto F and Vogel E 2009 *Physica* A **388** 2821
[59] Hartmann A K and Rieger H 2004 *New Optimization Algorithms in Physics* (Berlin: Wiley-VCH)
[60] Boettcher S and Percus A G 2001 *Phys. Rev. Lett.* **86**, 5211
[61] Dall J and Sibani P *Computer Physics Communications* **141** 260
[62] Ballesteros H G, Fernández L A, Martín-Mayor V, Muñoz Sudupe A, Parisi G and Ruiz-Lorenzo J J 1998

Phys. Rev. B **58** 2740
[63] Hasenbusch M *Phys. Rev.* B **82** 174433
[64] Pal K F 1996 *Physica* A **223** 283
[65] Hartmann A K 1997 *Europhys. Lett.* **40** 429

Journal of Physics: Conference Series **487** (2014) 012011 doi:10.1088/1742-6596/487/1/012011

Magnetic models on various topologies

F.W.S. Lima[1] **and J. A. Plascak**[1–3]

[1]Dietrich Stauffer Computational Physics Lab, Departamento de Física, Universidade Federal do Piauí , 64049-550, Teresina, PI, Brazil

[2]Departamento de Física, Universidade Federal de Minas Gerais, C. P. 702, 30123-970, Belo Horizonte, MG, Brazil

[3]Center for Simulational Physics, University of Georgia, 30602 Athens-GA, USA

E-mail: `fwslima@gmail.com,pla@hal.physast.uga.edu`

Abstract. A brief review is given on the study of the thermodynamic properties of spin models defined on different topologies like small-world, scale-free networks, random graphs and regular and random lattices. Ising, Potts and Blume-Capel models are considered. They are defined on complex lattices comprising Appolonian, Barabási-Albert, Voronoi-Delauny and small-world networks. The main emphasis is given on the corresponding phase transitions, transition temperatures, critical exponents and universality, compared to those obtained by the same models on regular Bravais lattices.

1. Introduction

The study and characterization of magnetic systems on regular d-dimensional lattices, both experimentally and theoretically, has been well established during the last century (see, for instance, the book series edited by Domb and Green, and also edited by Domb and Lebowitz, on *Phase Transitions and Critical Phenomena*[1, 2]). Although the transition temperatures are model dependent and non-universal, the critical exponents are known to be universal. In general, the universality class depends on the spatial lattice dimension d, on the number of components n (and symmetry) of the order parameter, and on the range of interactions. The agreement between theory and experimental results on real compounds has been reported to be excellent in this regard.

More recently, there has been a great deal of interest in studying networks, which are different from the regular crystalline Bravais lattices[3, 4, 5]. The study of such complex lattices, also called scale-free networks, has been mainly motivated by social organizations and computers connectivities[4], among others. It has been possible to recognize, in this way, networks ranging from networks in nature to networks of people as well.

Thus, it turns out quite interesting to understand the behavior of a magnetic system on such networks and also on random graphs. Despite being theoretical by now, this will certainly trigger the possibility of synthesizing experimental realizations of materials governed by random-lattices sites. Under this theoretical point of view, the magnetic model can be defined by usual spin interactions, where each spin is located on the sites of a complex lattice. The main questions that arise concern the possible existence of a first- or second-order (critical or multicritical) phase transition, and in the case of a second-order transition, the corresponding universality class of the model.

Table 1. Order of transition and critical exponents for the Ising and Potts models on regular lattices. In two dimensions we have the exact results and in three dimensions the values come from Monte Carlo simulations (see, for instance, reference[10]).

model	transition	β/ν	γ/ν	$1/\nu$
$2d$ Ising-S	2nd	0.125	1.75	1
$3d$ Ising-S	2nd	0.5185(15)	1.9630(30)	0.630(2)
$2d$ Potts-3	2nd	$2/15 = 0.133...$	$26/15 = 1.733...$	1.2
$2d$ Potts-4	2nd	0.125	1.75	1.5
$2d$ Potts > 4	1st	-	-	-

In this brief review we will consider complex lattices based on the Voronoi-Delauny (VD) tessellation[6], Barabási-Albert (BA)[3], Appolonian (AP)[7], and small-world (SW)[8] networks, and Erdös-Rènyi (ER)[9] random graphs. The magnetic models applied to these networks are the Ising, Potts and Blume-Capel models, among others. In the next section, a short introduction of the Hamiltonian models and lattices will be presented. In the following sections we will give a brief summary of what has been done on each network and random graphs and, in the final section, we will present a discussion and a summary of the results.

2. Models, lattices, networks and graphs

2.1. Models

The Ising and Blume-Capel models can be defined by the following Hamiltonian

$$\mathcal{H} = -J\sum_{\langle i,j\rangle}\sigma_i\sigma_j - H\sum_{i=1}^{N}\sigma_i + D\sum_{i=1}^{N}\sigma_i^2 , \qquad (1)$$

where the first sum is over nearest-neighbors on a d-dimensional lattice with N sites, σ_i is the state of a spin-S with components $\sigma_i = -S, -S+1, ...S-1, S$, and D is the crystal field. The general spin-S simple Ising model is recovered when $D = 0$. There is no transition for the one-dimensional model and in the two and three dimensions the universality class is defined by the exponents given in Tab. 1.

On the other hand, the q-state Potts model can be written as

$$\mathcal{H} = -J\sum_{\langle i,j\rangle}\delta_{\sigma_i\sigma_j} , \qquad (2)$$

where $\delta_{\sigma_i\sigma_j}$ is the Kronecker delta function and now $\sigma_i = 1, 2, ..., q$. Again, no transition is observed for the one-dimensional model. In two dimensions, which is the case treated in this work, the universality class is also given in Tab. 1 for $q \leq 4$ (for $q = 2$ it corresponds to the spin-1/2 Ising model). For $q > 4$ a first-order transition takes place.

The above models have also been treated in complex lattices. Some of these complex lattices will be briefly described below and more details can be found in references [3, 6, 7, 8, 9].

2.2. Directed and undirected Appolonian network

The Appolonian network is composed of $N = 3 + (3^n - 1)/2$ nodes, where n is the generation number and N the node number [7, 11]. On these AP structures we can introduce a disorder, in such a way that we redirect a fraction p of the links. This redirecting results in a directed network, preserving the outgoing node of the redirected link but changing the incoming node. When $p = 0$ we have the standard AP networks, while for $p = 1$ we have something similar to random networks [9]. In this procedure of the redirecting links, the number of outgoing links of each node is preserved even when $p = 1$ and the network still have hubs that are the most influent nodes. These networks display a scale-free degree distribution and a hierarchical structure. In the undirected case there exists the reciprocity of redirected link.

2.3. Directed and undirected small-world network
To generate the directed SW networks [8] we use a square grid and other irregular triangular. The disorder introduced here is identical to the procedure used on AP network cited above for both cases directed and undirected network.

2.4. Directed and undirected Erdös-Rènyi random graphs
An ER random graph is formed by a set of N vertices (sites) connected by K links (bonds)[9] . With a probability p a given pair of sites is connected by a bond type $p = 2K/N(N-1)$. The connectivity of a site is defined as the total number of bonds connected to it, like $k_i = \sum_j l_{ij}$, where $l_{ij} = 1$ if there is a link between the sites i and j and $l_{ij} = 0$ otherwise. Random graphs are completely characterized by the mean number of bonds per site, or the average connectivity $z = p(N-1)$. These links can be directed or undirected as well.

2.5. Directed and undirected Barabási-Albert Network
In the directed BA network, each new site added to the network selects (a preferential attachment proportional to the number of previous selections), with connectivity z, already existing sites or nodes as neighbors influencing it; the newly added site does not influence these neighbors. In the case of the undirected BA network, the newly added spin does influence these neighbors.

2.6. Directed and undirected Voronoi-Delaunay random lattices
In the undirected VD random lattice the construction of the lattice obey the following procedure: for each point in a given set of points in a plane, we determine the polygonal cell that contains the region of space nearest to that point than any other. Two cells are considered neighbors when they possess an extremity in common (Voronoi tessellation). From this Voronoi tessellation, we can obtain the dual lattice, or triangulation of Delaunay, by the following procedure. (i) When we have two neighbor cells, a link is placed between the two points located in the cells. (ii) From the links, we obtain the triangulation of space that is called the Delaunay lattice. (iii) The Delaunay lattice is dual to the Voronoi tessellation in the sense that its points correspond to cells, links to edges and triangles to the vertices of the Voronoi tessellation. The directed VD random lattices is constructed in the same way as the directed SW network.

3. Appolonian network

3.1. spin-1/2 Ising model
The Ising model has been studied on the triangular Appolonian networks with constant exchange interaction J [11, 12] and with node-dependent interaction constants $J_{i,j}$ [11, 13]. The corresponding thermodynamic and magnetic properties have been obtained by formulating the problem in terms of transfer matrices. In both cases the results show no evidence of any phase transition either for the ferromagnetic or the antiferromagnetic model, as well as for short or long ranged interactions. Quenched random models, including ferromagnetic and antiferromagnetic bond dilution and Ising spin glasses have already been considered [14]. In this case the ordered phases persist up to infinite temperature over the entire range of disorder.

3.2. q-state Potts model
The q-state Potts model has been studied on Appolonian networks by using Monte Carlo simulations and transfer matrix techniques[15]. In this case, as for the spin-1/2 Ising model, no transition has been detected for any value of q.

4. Erdös-Rènyi random graphs
4.1. Ising model
The spin-1/2 and spin-1 Ising model have been studied on directed and undirected Erdös-Rènyi random graphs with different coordination numbers by using Monte Carlo simulations[16]. For spin-1/2, the model exhibits a critical temperature, below which there is a spontaneous magnetization, with mean-field exponents. Although the critical temperature depends on being directed or undirected graphs, the critical exponents are the same as the mean-field ones for both cases. The same qualitative results are also obtained for the spin-1 model. However, in this case, one has a first-order phase transition instead. These results are in agreement with those obtained for a non equilibrium model defined on the same random Erdös-Rènyi lattices [17, 18] (in this case, according to the Grinstein, Jayaprakash and He criterion [19], non-nequilibrium and equilibrium models that present up-down symmetry may belong to the same universality class).

5. Barabási-Albert Network
5.1. Ising model
The spin-1/2 Ising model was first studied on undirected scale-free Barabási-Albert network (UBA) by Aleksiejuk *et al.* [20]. Through Monte Carlo simulations [20] they have shown that there exists a Curie temperature which logarithmically increases with the increasing of the system size N. After, Sumour *et al.* [21, 22] have treated the Ising model on a directed Barabási-Albert network (DBA) using standard Glauber kinetic Ising model to a fixed network. Unlike the previous results presented by Aleksiejuk *et al.* [20], they have shown that the spin-1/2 Ising model on DBA does not exhibit any phase transition for temperatures different from zero and confirmed that the corresponding Curie temperature presents an asymptotic Arrhenius law extrapolation for the relaxation time τ (defined as the first time when the sign of the magnetization flips) of the type $1/\ln\tau \propto T$. This means in fact that at all finite temperatures the magnetization eventually vanishes, i.e., no ferromagnetism is present. Lima et al. [23] have also studied the Ising model with spin S = 1, 3/2 and 2 on DBA and they have shown that no phase transition is present on these DBA networks.

5.2. q-state Potts model
The Potts model with $q = 3$ and 8 states have been studied on DBA through Monte Carlo simulations by Lima [24]. Surprisingly, different from the conventional results for the Potts model on regular two-dimensional lattices, the Potts model on DBA presented a phase transition of first order for both values of $q = 3$ and 8. Sumuor and Lima [25] have studied the critical behavior of Ising and Potts models on semi-directed Barabási-Albert network (SDBA), recently studied by Sumour and Radwan [26], where now the number $N(k)$ of nodes with k links each decays as $1/k^{\gamma}$ and the exponent γ decreases from 3 to 2 for increasing m, where m is the number of old nodes which a new node added to the network selects to be connected with. This behavior is totally different from UBA and DBA scale free networks where $\gamma = 3$ is universal, i.e., independent of m. For both Ising and Potts model the results showed no phase transition, in agreement with references[21, 22], and a Curie temperature to diverge at positive temperatures $T_c(N)$ by a Vogel-Fulcher law, with $T_c(N)$ increasing logarithmically with network size N.

6. Small-World Network

6.1. Ising model

The one-dimensional spin-1/2 Ising model has been studied, via Monte Carlo simulations, on small-world network, where the small-world connections have a form proportional to $r_{ij}^{-\alpha}$. It has been shown that any non-zero value of α destroys the finite-temperature transition in the thermodynamic limit[27]. The results are different when the small-world interactions are independent of the length ($\alpha = 0$), where a finite-temperature phase transition is observed [28, 29, 30, 31, 32, 33] with critical exponents smaller than the corresponding ones for the two-dimensional Ising model.

The two- [34, 35] and three-dimensional spin-1/2 model have also been treated by Monte Carlo simulations [36]. It has been shown that in the thermodynamic limit the phase transition has a mean-field character.

On the other hand, the two-dimensional model has been extended for greater values of spin, namely spin-1, 3/2 and 2 on directed small-world networks (DSW) [23]. In the DSW network [37], Lima *et al.* [23] showed that the two-dimensional Ising model exhibits a second and first-order phase transition for rewiring probability $p = 0.2$ and 0.8, respectively, for spin values $S = 1, 3/2$ and 2. For values $p > p_c \sim 0.25$ this model presents a first-order phase transition[38]. These results are, nevertheless, in accordance with the those from Sánchez *et al.* [37] for the corresponding non-equilibrium model.

6.2. q-state Potts model

Recently, Silva *et al.* [39], through Monte Carlo simulations, have studied the two-dimensional Potts models with $q = 3$ and 4 states on DSW network. They have found that this model exhibits a first-order and a second-order phase transition for $q = 3$, depending on the rewiring probability p values. However, for $q = 4$ the system presents only a first-order phase transition for any value of p different of zero. This critical behavior on DSW is different from the Potts model on a regular square lattice, where the second-order phase transition is present for $q \leq 4$ and a first-order phase transition takes place for $q > 4$.

7. Voronoi-Delaunay random lattices

7.1. Ising model

The spin-1/2 Ising model was first studied, via Monte Carlo simulations, on two-dimensional Voronoi-Delaunay random lattices by Espriu *et al.* [40]. Their results, obtained with Metropolis update algorithm, showed weak evidence that the critical exponents belong to the same universality class of the spin-1/2 Ising model on regular lattices in two-dimensions. Janke *et al* [41, 42], using single-cluster Monte Carlo update algorithm [43], reweighing techniques [44] and finite size scaling analysis, also studied the Ising model on the two-dimensional Voronoi-Delaunay random lattices and their results confirmed the initial evidence obtained by Espriu *et al.* [40], namely, that the Ising model on these random lattices belong in fact to the same universality class of the Ising model in regular two-dimensional lattices. Afterwards, Janke *et al* [45, 46], using the same algorithm earlier applied for the two-dimensional Ising model, studied the three-dimensional version on the VD random lattices. Again, their results showed that the system belongs to the same universality class of the Ising model in three-dimensional regular lattices. These new results, however, seem to be in disagreement with Harris criterion [47]. On the other hand, Lima *et al.* [48, 49] have also studied this model by assuming that the exchange coupling J varies with the distance r between the first neighbors as $J(r) \propto e^{-\alpha r}$, with $\alpha \geq 0$. The model in two and three dimensions was simulated applying the single-cluster Monte Carlo update algorithm and the reweighing technique. The results also showed that this random system belongs to the same universality class as the pure two and three-dimensional ferromagnetic Ising model. In addition, within the context of persistence, Lima *et al* [50] have studied the Ising model on VD random lattices using the zero-temperature Glauber dynamics. They showed that the model has exhibited blocking effect, which means that the persistence does not go to zero when time $t \to \infty$.

Models with spin greater than 1/2 have also been treated by Monte Carlo simulations. While for the spin-1 Ising model one also gets the same universality class as for the model on regular lattices [51], the same is not true for the spin-3/2 case, where the critical exponents are different[52].

The spin-1 Blume-Capel was treated on directed Voronoi-Delauny lattices[51] and, as in the directed small-world case above, a continuous transition has been observed only for rewiring probabilities $p < p_c \sim 0.35$, while for $p > p_c$ a first-order transition takes place.

7.2. q-state Potts model

Janke *et al.* [53] have treated the Potts model with $q = 8$ states on two-dimensional VD random lattices using the same algorithm and technique from references [45, 46]. They have obtained the same first-order behavior as for the model on a regular square lattice. It is worth to stress that Chen *et al.* [54] have treated the quenched bond randomness of this model on regular square lattice and have obtained a phase transition which changed from first to second order. Afterwards, Lima *et al.* [55] have also studied the $q = 8$ Potts model on VD random lattices using the same coupling factor $J(r) \propto e^{-\alpha r}$ from references [48, 49]. Using Monte Carlo simulations, and the same procedure from references [45, 46], they have shown that for $\alpha = 0$ only first-order transition is present, in agreement with previous works [53]. However, for $\alpha > 0$ the $q = 8$ Potts model on VD random lattices displays a second-order transition with the critical exponents β/ν and γ/ν different from the Ising model on regular square lattices. For the $q = 3$ Potts model[56], the critical exponents are different from those of the model on a two-dimensional regular lattice, however, the exponents ratio β/ν and γ/ν seem to be identical to the corresponding ratio on regular lattices.

Recently, Lima [57] has studied the $q = 4, 6$ and 8 Potts model. The $q = 4$ model exhibits a second-order transition for $\alpha = 0$, 0.5 and 1 with critical exponents that are α dependent, and different from those exponents on regular lattices. For $q = 6$ and $q = 8$, the system undergoes a first-order transition for $\alpha = 0$ and a second-order transition for $\alpha > 0$, with critical exponents again dependent on α and different from those on regular lattices.

7.3. Harris criterion and Voronoi-Delaunay random lattices

The Harris's criterion[47] for regular lattices (for instance, square, triangular, cubic and hyper-cubic lattices) is based on the exponent of the specific heat of a pure system (free of impurities), to determine whether or not the system will change its critical behavior or universality class in the presence of impurities or some kind of topological disorder. Since then, it has been well established that when $\alpha < 0$ (not to be confused with the exchange decay exponent used in the previous sections) the critical behavior of the specific heat is unchanged when impurities are added to the magnetic system in a random way. On the other hand, for $\alpha > 0$ the system with randomness has a critical behavior different from the pure system case. However, for the marginal case, i.e, $\alpha = 0$ (two-dimensional Ising model) the Harris's criterion predict that both earlier behavior can occur with a pure system in the presence of some type of disorder (for the case $\alpha = 0$ see, for instance, [58, 59] and references therein). However, as mentioned previously, Janke *et al*[41, 42, 45, 46] and Lima *et al* [48, 49] showed that the Harris's criterion fails on two and three-dimensional Ising model on the VD random lattices.

In order to better understand the Harris's criterion on the VD random lattices Janke *et al.* [60] investigated the applicability of this criterion to the case of spin models on VD random lattices. They were interested in verifying whether the critical behavior depends on the degree of spatial correlations present in the random lattices, which was quantified by a *wandering exponent w* . They determined the numerical value of *wandering exponent* $w = 0.5$ for both Ising and Potts models. However, this value $w = 0.5$ corresponds to case $\alpha = 0$ of the Harris's criterion and again, for these spin models on VD random lattices, nothing can be predicted about the corresponding critical behavior of magnetic models.

8. Discussion

Continuous symmetry spin models have recently been treated on complex topologies such as the XY model on dilute Lévi graphs [61]. We should stress that other non-magnetic systems have also been studied on such complex networks, for instance, the majority vote model on Archimedian lattices (see references[62, 63] and references therein) and the non-equilibrium contact-process on Voronoi-Delauny lattice[64], which are not being treated here. In Table 2 we try to summarize the main results of the magnetic models with discrete degree of freedom treated on random lattices. For the readers convenience, we have also in Table 1 the transition and critical exponents of the same models on regular lattices.

From Table 2 one can see that there is, in some cases, a change in the order of the transition and, in others, no transition at all. When the transition remains continuous, in general the exponents also change to another universality class, except for the spin-1/2 and spin-1 Ising model on the Voronoi-Delauny lattice and the spin-1 Ising model on the directed VD lattice, where the results for the exponents give an indication of belonging to the same universality class of the corresponding model on regular latices. It is also clear that the Harris criterion seems not to be effective for these random lattices, because one cannot say a priori either what the order of the transition will be or the universality class of the possible second-order transition. One can argue, of course, that the simulations are still not in the true finite size scaling regime and a change on the results could be obtained by considering larger lattices. However, from the quality of the results shown in the simulations it seems not so plausible such an idea. Thus, apart from some lacking models still to be treated in order to complete Table 2, the question of the order of the transition and the corresponding universality class of the model defined on complex topologies seem to be an open question still to be answered in more general terms

9. Acknowledgments

The authors would like to thank M. Novotny, R. F. S. Andrade and N. M. de Araujo for a critical reading of the manuscript. Financial support from CNPq, CAPES, FAPEMIG, and FAPEPI are also gratefully acknowledged.

References

[1] Domb C and Green M S 1972-1976 *Phase Transition and Critical Phenomena* Eds. Vols. 1-6 (London: Academic Press)
[2] Domb C and Lebowitz J L 1983-2000 *Phase Transition and Critical Phenomena* Eds. Vols. 7-20 (London: Academic Press)
[3] Barabási A-L and Albert R 1999 *Science* **286** 509
[4] Albert R and Barabási A-L 2002 *Rev. Mod. Phys.* **74** 47
[5] Barabási A-L 2009 *Science* **325** 412
[6] Christ N R, Friedberg R and Lee T D 1982 *Nucl. Phys. B* **202** 89
[7] Boyd D W 1973 *Can. J. Phys.* **25** 303; Herrmann H J, Mantica G and Bessis D 1990 *Phys. Rev. Lett.* **65** 3223
[8] Watts D J and Strogatz S H 1998 *Nature* **393** 440
[9] Erdös P and Rènyi A 1960 *Publ. Math. Inst. Hung. Acad. Sci.* **5** 17
[10] Landau D P and Binder K 2009 *A Guide to Monte Carlo Simulations in Statistical Physics* (New York: Cambridge University Press)
[11] Andrade R F S and Herrmann H J 2005 *Phys. Rev. E* **71** 056131
[12] Andrade Jr. J S, Herrmann H J, Andrade R F S and da Silva L R 2005 *Phys. Rev. Lett.* **94** 018702
[13] Andrade R F S, J S Andrade Jr, and Herrmann H J 2009 *Phys. Rev. E* **79** 036105
[14] Nadir Kaplan C, Hinczewski M and Nihat Berker A 2009 *Phys. Rev. E* **79** 061120

Journal of Physics: Conference Series **487** (2014) 012011 doi:10.1088/1742-6596/487/1/012011

Table 2. Order of transition and critical exponents (when not available it is indicated by dots) for different models on different random lattices. S stands for the value of the Ising spin and q for the number of states of the Potts model (or Ising-S and Potts-q). d stands for the dimension of the lattice, and p for the rewiring probability of the corresponding directed random lattice. α gives the range of the interaction through $J(r) \propto e^{-\alpha r}$ (unless stated $\alpha = 0$). MF - mean-field.

model	transition	β/ν	γ/ν	$1/\nu$	universality	Ref.
			Appolonian			
Ising-1/2	no	-	-	-	-	[11]
Potts	no	-	-	-	-	[13]
			Erdös-Rènyi			
Ising-1/2	2nd	1	2	2	MF	[16, 17, 18]
Ising-1	1st	-	-	-	-	[17, 18]
			Barabási-Albert			
Ising-S	no	-	-	-	-	[21, 22, 23]
Potts-q	no	-	-	-	-	
			Small World			
$1d$ Ising-1/2	2nd	~ 0.0001	0.6(1)	...		[28]-[33]
$2, 3d$ Ising-1/2	2nd	1	2	2	MF	[36]
			Ising on directed SW			
S-1/2 any p	2nd	0.44(3)	1.129(6)	1.14(2)	...	[38]
S-1, $p < 0.25$	2nd	0.40(3)	1.22(3)	1.16(5)	...	[38]
S-1, $p > 0.25$	1st	-	-	-	-	[38]
			Potts on directed SW			
q=3 p=0.1	2nd	0.24(5)	1.5(1)	...	...	[39]
q=3 p=0.9	1st	-	-	-	-	[39]
q=4 any p	1st	-	-	-	-	[39]
			Voronoi-Delauny			
			2d Ising			
S-1/2	2nd	0.1208(92)	1.7503(59)	0.964(28)	Ising	[42]
S-1	2nd	0.135(9)	1.751(4)	1.016(8)	Ising	[51]
S-3/2	2nd	0.331(9)	1.467(9)	1.13(3)	...	[52]
			3d Ising			
S-1/2	2nd	0.51587(82)	1.9576(13)	1.5875(12)	Ising	[45]
			2d Potts			
q=3 α=0	2nd	0.133(13)	1.764(18)	1.19(2)	...	[56]
q=3 α=0.5	2nd	0.118(12)	1.751(17)	1.07(1)	...	[56]
q=3 α=1	2nd	0.106(12)	1.754(18)	0.94(1)	...	[56]
q=4 α=0	2nd	0.143(9)	1.799(6)	1.377(9)	...	[57]
q=4 α=0.5	2nd	0.12(2)	1.70(2)	1.13(2)	...	[57]
q=4 α=1	2nd	0.10(2)	1.66(4)	0.93(3)	...	[57]
q=6, 8 α=0	1st	-	-	-	-	[53, 55, 57]
q=6 α=0.5	2nd	0.122(4)	1.53(5)	1.09(1)	...	[57]
q=6 α=1	2nd	0.14(1)	1.56(5)	0.91(1)		...[57]
q=8 α=0.5	2nd	0.131(7)	1.20(8)	0.83(3)	...	[57]
q=8 α=1	2nd	0.126(8)	1.45(6)	0.79(3)	...	[57]
			2d Ising-1 on directed VD			
$p < 0.35$	2nd	0.421(7)	1.101(4)	1.105(8)	Ising	[51]
$p > 0.35$	1st	-	-	-	-	[51]

[15] Araújo N A M, Andrade R F S and Herrmann H J 2010 *Phys. Rev. E* **82** 046109
[16] Lima F W S and Sumour M A 2012 *Physica A* **391** 948
[17] Pereira L F C and Brady Moreira F G 2005 *Phys. Rev. E* **71** 016123
[18] Lima F W S, Sousa A O and Sumour M A 2008 *Physica A* **387** 3503
[19] Grinstein G, Jayaprakash C and He Y 1985 *Phys. Rev. Lett.* **55** 2527
[20] Aleksiejuk A, Hołyst J A and Stauffer D 2002 *Physica A* **310** 269
[21] Sumour M A and Shabat M M 2005 *Int. J. Mod. Phys. C* **16** 585
[22] Sumour M A, Shabat M M and Stauffer D 2006 *Islamic University Journal (Gaza)* **14** 209; cond-mat/0504460
[23] Lima F W S, Luz M S E and Costa Filho R N 2009 *Multidiscipline Modeling in Mat. and Str.* **5** 223
[24] Lima F W S 2007 *Comm. Comp. Phys.* **2** 522
[25] Sumour M A and Lima F W S 2013 *submitted to Comp. Phys. Comm.*
[26] Sumour M A and Radwan M A 2013 *Int. J. Mod. Phys. C* **23** 1250062
[27] Jeong D, Hong H, Kim B J and Kim M Y 2003 *Phys. Rev. E* **68** 027101
[28] Gitterman M 2000 *J. Phys. A: Mathe. Gen.* **33** 8373
[29] Barrat A and Weight M 2000 *Eur. Phys. J. B* **13** 547
[30] Pekalski A 2001 *Phys. Rev. E* **64** 057104
[31] Hong H, Kim B J and Choi M Y 2002 *Phys. Rev. E* **66** 018101
[32] Kim B J, Hong H, Holme P, Jeon GS, Minnhagen P and Choi M Y 2001 *Phys. Rev. E* **64** 056135
[33] Novotny M A and Wheeler S M 2004 *Braz. J. Phys.* **34** 395
[34] Zingjun Zhang 2005 *Critical Properties of Small World Ising Models* (PhD Dissertation, Mississippi State University). Available on the web at http://sun.library.msstate.edu/ETD-db/theses/available/etd-11102005-220554/
[35] Novotny M A et al. 2005 *J. Appl. Phys.* **97** 10B309
[36] Herrero C P 2002 *Phys. Rev. E* **65** 066110
[37] Sánchez A D, López J M and Rodriguez M A 2002 *Phys. Rev. Lett.* **88** 048701
[38] Lima F W S and Plascak J A 2013 *Eur. Phys. J. B* **86** 300
[39] da Silva P R O, Lima F W S and Costa Filho R N 2013 *in press Comp. Phys. Comm.*
[40] Espriu M G, Rakow P E L and Wheater J F 1986 *Nucl. Phys. B* **256** 92
[41] Janke W, Katoot M and Villanova R 1993 Phys. Lett. B **315** 412
[42] Janke W, Katoot M and Villanova R 1994 Phys. Rev. B **49** 9644
[43] Wolff U 1989 *Phys. Rev. Lett.* **62** 361
[44] Ferrenberg A M and Swendsen R H 1988 *Phys. Rev. Lett.* **61** 2635
[45] Janke W, Katoot M and Villanova R 2000 Nucl. Phys. B **83-84** 697
[46] Janke W and Villanova R 2002 Phys. Rev. B **66** 134208
[47] Harris A B 1974 *J. Phys. C* **7** 1671
[48] Lima F W S, Moreira J E, J S Andrade Jr and Costa U M S 2000 *Physica A* **283** 100
[49] Lima F W S, Costa Filho R N and Costa U M S 2008 *Physica A* **287** 1545
[50] Lima F W S, Costa Filho R N and Costa U M S 2004 *J. of Mag. Mag. Mat.* **270** 182
[51] Fernandes F P, Lima F W S and Plascak J A 2010 *Comp. Phys. Comm.* **181** 1218
[52] Lima F W S and Plascak J A 2006 *Braz. J. Phys.* **36** 660
[53] Janke W and Villanova R 1995 *Phys. Lett. A* **209** 179
[54] Chen S, Ferrenberg A M and Landau D P 1992 *Phys. Rev. Lett.* **69** 1213
[55] Lima F W S, Moreira J E, J S Andrade Jr and Costa U M S 2000 *Eur. Phys. J. B* **13**, 107
[56] Lima F W S, Costa U M S, Almeida M P and Andrade Jr J S 2000 *Eur. Phys. J. B* **17**, 111
[57] Lima F W S 2010 *Physica A* **389** 3039
[58] Novotny M A 1993 *Phys. Rev. Lett.* **70** 109
[59] Martins P H L and Plascak J A 2007 *Phys. Rev. E* **76** 012102
[60] Janke W and Weigel M 2004 *Phys. Rev. B* **69** 144208.
[61] Berganza M I and Leuzzi L 2013 *Phys. Rev. B* **88** 144104.
[62] Lima F W S, Fukco U L and Costa Filho R N 2005 *Phys. Rev. E* **71** 036105
[63] Lima F W S 2013 submitted to Eur. Phys. J. P.
[64] de Oliveira M M, Alves S G, Ferreira S C, and Dickman R 2008 *Phys. Rev. E* **78** 031133

Journal of Physics: Conference Series **487** (2014) 012012 doi:10.1088/1742-6596/487/1/012012

Simulating Thin Sheets: Buckling, Wrinkling, Folding and Growth

Roman Vetter[1], **Norbert Stoop**[1,2], **Falk K. Wittel**[1], **Hans J. Herrmann**[1]

[1]Computational Physics for Engineering Materials, IfB, ETH Zurich, Schafmattstrasse 6, CH-8093 Zurich, Switzerland

[2]IAS Institute of Applied Simulations, ZHAW Zurich University of Applied Sciences, CH-8820 Wädenswil, Switzerland

E-mail: `vetterro@ethz.ch`

Abstract. Numerical simulations of thin sheets undergoing large deformations are computationally challenging. Depending on the scenario, they may spontaneously buckle, wrinkle, fold, or crumple. Nature's thin tissues often experience significant anisotropic growth, which can act as the driving force for such instabilities. We use a recently developed finite element model to simulate the rich variety of nonlinear responses of Kirchhoff-Love sheets. The model uses subdivision surface shape functions in order to guarantee convergence of the method, and to allow a finite element description of anisotropically growing sheets in the classical Rayleigh–Ritz formalism. We illustrate the great potential in this approach by simulating the inflation of airbags, the buckling of a stretched cylinder, as well as the formation and scaling of wrinkles at free boundaries of growing sheets. Finally, we compare the folding of spatially confined sheets subject to growth and shrinking confinement to find that the two processes are equivalent.

1. Introduction

Thin sheets are omnipresent in nature, technology and everyday life, appearing at virtually all length scales. Being much thinner in one than in the other two dimensions, they can develop an unparalleled, rich variety of deformation modes when subjected to external forces, spatial constraints, or intrinsic growth. They buckle, wrinkle, fold, and crumple. Numerical simulations are an often-indispensable approach for studying the complex interplay of these modes. The folding and crumpling of a piece of paper [1–7] and metal sheets wrinkling and crumpling in vehicle collisions [8–11] are two out of many examples. For many such problems, the finite element method (FEM) has shown to be amongst the most efficient and flexible tools, especially in cases with strong material nonlinearity, complex geometry, or anisotropy. Even though the Kirchhoff–Love theory [12] provides a simple kinematic description, numerically sound finite element implementations of thin sheets have turned out to be difficult and cumbersome in the past. These problems can be successfully overcome since the subdivision surface paradigm was introduced to the FEM [13,14].

Large deformations of soft thin tissue such as insect wings, plant leaves, cell membranes, or flowers are often induced by growth (or shrinkage) [15,16], inevitably leading to the development of residual stress [17,18]. In this paper, we present an extension of the Kirchhoff–Love theory to allow for anisotropic in-plane growth, which we implement with Loop subdivision shape functions. The combination of these two concepts grants access to a very straightforward and highly efficient, yet powerful and flexible numerical tool for the simulation of nonlinear thin sheet mechanics. Our approach accounts for the change of reference curvature when the surface grows, generalizing recently developed tethered mass-spring models [19, 20]. The next section summarizes the mentioned extension. It is followed in the subsequent sections by a series of thin sheet problems that we solve using the developed FEM implementation. Special attention is paid to the formation of self-similar wrinkles along a plastically stretched free edge as well as on the scaling of single-wavelength wrinkles of growing cylinders similar to flowers.

2. The Kirchhoff-Love Sheet with Anisotropic Growth

Let $\overline{\Omega} \subset \mathbb{E}^3$ be the stress-free undeformed ("reference") middle surface of a sheet with small thickness h. Under the action of external forces or growth, the sheet deforms into a new configuration with middle surface $\Omega \subset \mathbb{E}^3$. In the following, let Greek indices $\alpha, \beta, \gamma, \delta \in \{1, 2\}$, and Latin indices $i, j \in \{1, 2, 3\}$. Lower (upper) indices will denote covariant (contravariant) components. Moreover, let $\{\theta^1, \theta^2, \theta^3\}$ be a curvilinear coordinate system, and let $\overline{\mathbf{x}}(\theta^1, \theta^2)$ and $\mathbf{x}(\theta^1, \theta^2)$ be parametrizations of $\overline{\Omega}$ and Ω, respectively (see Fig. 1). The material points $\overline{\mathbf{p}}$ and $\mathbf{p} = \chi(\overline{\mathbf{p}})$ in the reference and deformed sheet are parametrized as

$$\overline{\mathbf{p}}(\theta^1, \theta^2, \theta^3) = \overline{\mathbf{x}}(\theta^1, \theta^2) + \theta^3 \overline{\mathbf{a}}_3(\theta^1, \theta^2) \quad \text{and} \quad \mathbf{p}(\theta^1, \theta^2, \theta^3) = \mathbf{x}(\theta^1, \theta^2) + \theta^3 \mathbf{a}_3(\theta^1, \theta^2), \tag{1}$$

where $\theta^3 \in [-h/2, h/2]$. χ is a diffeomorphism that maps from the reference to the deformed material positions. The tangent spaces of $\overline{\Omega}$ and Ω are spanned by the respective vector fields

$$\overline{\mathbf{a}}_\alpha(\theta^1, \theta^2) = \overline{\mathbf{x}}_{,\alpha} = \frac{\partial \overline{\mathbf{x}}}{\partial \theta^\alpha} \quad \text{and} \quad \mathbf{a}_\alpha(\theta^1, \theta^2) = \mathbf{x}_{,\alpha} = \frac{\partial \mathbf{x}}{\partial \theta^\alpha}. \tag{2}$$

By virtue of the Kirchhoff assumption, straight material lines normal to the middle surface retain these properties as well as their length. They are determined by the unit normal vectors

$$\overline{\mathbf{a}}_3 = \frac{\overline{\mathbf{a}}_1 \times \overline{\mathbf{a}}_2}{|\overline{\mathbf{a}}_1 \times \overline{\mathbf{a}}_2|} \quad \text{and} \quad \mathbf{a}_3 = \frac{\mathbf{a}_1 \times \mathbf{a}_2}{|\mathbf{a}_1 \times \mathbf{a}_2|}. \tag{3}$$

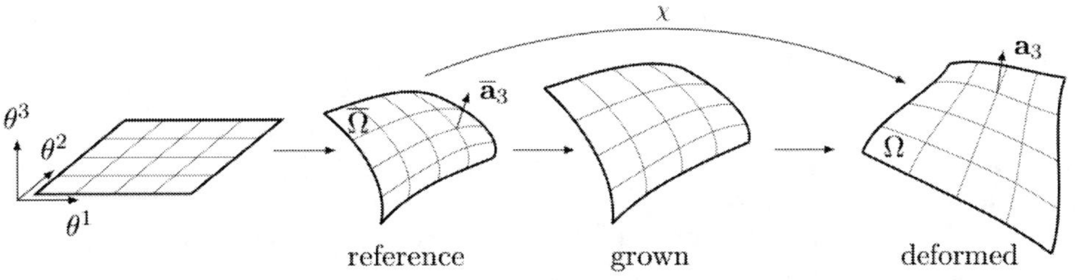

Figure 1. Reference, grown and deformed configurations of the sheet's middle surface.

The covariant components of the first fundamental forms follow as

$$\overline{a}_{\alpha\beta} = \overline{\mathbf{a}}_\alpha \cdot \overline{\mathbf{a}}_\beta \quad \text{and} \quad a_{\alpha\beta} = \mathbf{a}_\alpha \cdot \mathbf{a}_\beta, \tag{4}$$

while those of the second fundamental forms are given by

$$\overline{b}_{\alpha\beta} = \overline{\mathbf{a}}_3 \cdot \overline{\mathbf{a}}_{\alpha,\beta} \quad \text{and} \quad b_{\alpha\beta} = \mathbf{a}_3 \cdot \mathbf{a}_{\alpha,\beta}. \tag{5}$$

Assuming that the thin sheet obeys the St. Venant-Kirchhoff law of linear elasticity, the connection between its kinematics and energetics is provided by the Koiter energy density functional [21,22]. Let the sheet be characterized by Young's modulus E and Poisson's ratio ν. The elastic energy U_e of the Koiter sheet is obtained by integrating the energy per unit area over the middle surface:

$$U_e[\bar{\mathbf{x}}, \mathbf{x}] = \frac{1}{2} \int_{\bar{\Omega}} \frac{Eh}{1-\nu^2} \left(H^{\alpha\beta\gamma\delta} \alpha_{\alpha\beta} \alpha_{\gamma\delta} + \frac{h^2}{12} H^{\alpha\beta\gamma\delta} \beta_{\alpha\beta} \beta_{\gamma\delta} \right) \mathrm{d}\bar{\Omega}, \tag{6}$$

where $\mathrm{d}\bar{\Omega} = |\bar{\mathbf{a}}_1 \times \bar{\mathbf{a}}_2| \, \mathrm{d}\theta^1 \mathrm{d}\theta^2$. The Einstein summation applies to repeated indices. H is often referred to as the "elastic tensor", and is given component-wise by

$$H^{\alpha\beta\gamma\delta} = \nu \bar{a}^{\alpha\beta} \bar{a}^{\gamma\delta} + \frac{1-\nu}{2} (\bar{a}^{\alpha\gamma} \bar{a}^{\beta\delta} + \bar{a}^{\alpha\delta} \bar{a}^{\beta\gamma}). \tag{7}$$

$\alpha = (a - \bar{a})/2$ and $\beta = \bar{b} - b$ are the in-plane (2×2) membrane and bending strain tensors, respectively. The Koiter shell (6) can be extended to incorporate anisotropic growth through the multiplicative decomposition of the geometric deformation gradient $\nabla\chi = \mathbf{F}_e \mathbf{F}_g$ [23,24] into a growth tensor $\mathbf{F}_g$ and a purely elastic response $\mathbf{F}_e$, that ensures continuity and compatibility of the body. Owing to the Kirchhoff constraints, we may write

$$\mathbf{F}_g = \begin{bmatrix} G & 0 \\ 0^T & 1 \end{bmatrix} \qquad (G \in \mathbb{R}^{2\times 2} \text{ symmetric}), \tag{8}$$

and the growth-modified elastic strains then simply read [25]

$$\alpha = \frac{1}{2} \left(G^{-T} a G^{-1} - \bar{a} \right), \tag{9}$$

$$\beta = \bar{b} - G^{-T} b G^{-1}. \tag{10}$$

We further augment the elastic energy (6) with an inertial term to capture the dynamics of the thin sheet. The kinetic energy reads

$$U_k[\bar{\mathbf{x}}, \mathbf{x}] = \frac{1}{2} \int_{\bar{\Omega}} h\rho \, \dot{\mathbf{x}} \cdot \dot{\mathbf{x}} \, \mathrm{d}\bar{\Omega}, \tag{11}$$

where ρ is the mass density of the sheet, and $\dot{\mathbf{x}} = \partial \mathbf{x}/\partial t$ is the velocity. Our aim is to find the minimizer $\mathbf{x}$ of the total energy $U = U_e + U_k$ for given growth tensors $\mathbf{F}_g$ or external driving forces.

3. Finite Element Implementation with Subdivision Surfaces

To account for out-of-plane bending rigidity, the bending term in Eq. (6) integrates the Gaussian and mean curvatures, which comprise second derivatives of the displacement field $\mathbf{u} = \mathbf{x} - \bar{\mathbf{x}}$, over the middle surface. For boundedness of the integral in the weak formulation, continuously differentiable finite element shape functions (C^1-continuity) are needed. This requirement has proven very challenging in the history of shell finite elements, until Cirak et al. [13,14] have introduced Loop subdivision surfaces to the FEM. A fundamental difference to traditional finite elements is that subdivision surfaces gain C^1-continuity at the expense of a larger support of the individual shape functions. Details on their implementation are given in Refs. [13,25].

Aside from guaranteeing convergence, subdivision surfaces allow a classical Rayleigh–Ritz formulation of the sought finite element deformation: no rotational variables are needed and the only unknowns are the three nodal displacements. Moreover, a single quadrature point per element is sufficient [13, 14, 25], rendering this finite element approach computationally highly efficient and flexible. Of course, increasing the number of quadrature points may assist in resolving strongly anisotropic growth fields or constitutive relations. We employ Loop subdivision surface shape functions here to minimize the total energy U numerically by solving Newton's equation of motion with a standard predictor-corrector scheme. Subcritical viscous damping is added for numerical stability and equilibration.

4. Inflated Pillows and Airbags

As a first instance of folding and wrinkling, we reproduce the inflation of pillows and airbags from Ref. [14]. A square sheet with diagonal length $d = 120$ cm and thickness $h = 1$ mm and a circular sheet with radius $R = 35$ cm and thickness $h = 0.4$ mm are instantaneously pressurized with 5 kPa to buckle out of their flat initial configuration. The elastic moduli are given by $E = 588$ MPa, $\nu = 0.4$ and $E = 60$ MPa, $\nu = 0.3$, respectively. We exploit the reflection symmetry by simulating only the upper half of the geometry. The quasi-static equilibrium configurations are shown in Fig. 2, and a movie of the dynamic deformation is provided in the supplementary material. The square pillow features a distinct folding pattern in the middle of the four edges, resulting from the non-uniform distribution of Gaussian curvature: The edges are closer to the point of maximal uplift than the corners, thus getting curved more and pulled towards the center. They are laterally stretched and longitudinally compressed, yielding the observed folds. The circular airbag, on the other hand, is initially axisymmetric and therefore behaves differently. Wrinkles develop similarly to elastic plates stamped into curved cavities [26] or ultrathin films placed on fluid drops [27]. On top of these low-amplitude wrinkles, axisymmetry of the stress field is broken and large crumples occur that localize the geometrically imposed Gaussian curvature.

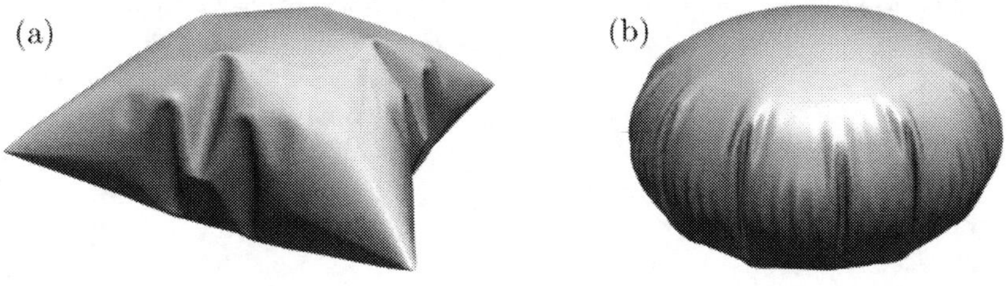

Figure 2. Inflated square pillow (a) and circular airbag (b).

5. Buckling of a Stretched Cylinder

A frequently studied buckling problem is the laterally stretched open cylinder [28, 29]. Two opposite point forces of equal increasing magnitude F cause the cylindrical sheet with radius $R = 4.953$ cm, length $L = 10.35$ cm, thickness $h = 0.94$ mm, and free edges to first bend before snapping through at $F_c = 11.836\, Eh^3/R$ to the post-buckling regime, where further deformations are dominated by stretching. The elastic moduli a fixed to $E = 10.5$ MPa, $\nu = 5/16$. In Fig. 3, we plot the radial displacements of the points A, B and C, which are degenerate at the buckling threshold F_c. The corresponding movie can be found in the supplementary material.

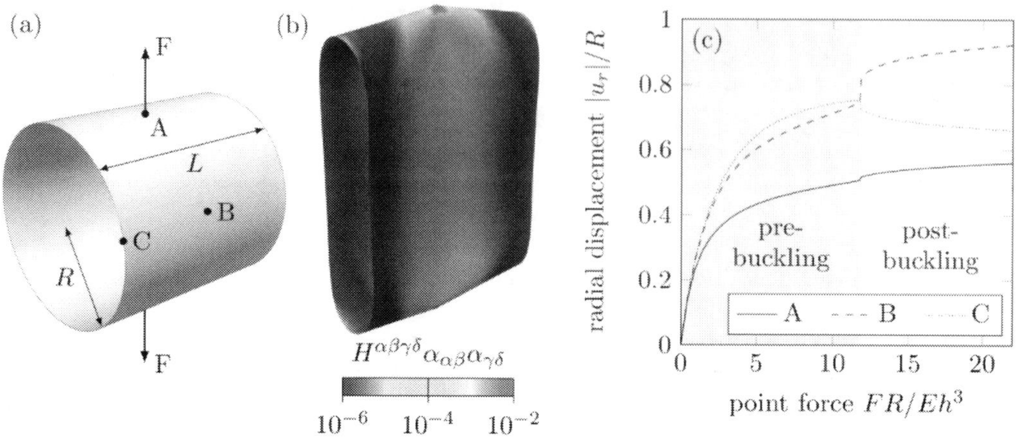

Figure 3. (a) Reference state of the cylindrical sheet. (b) Equilibrium solution at $F = 1.92F_c$. The stretching energy density is shown on a logarithmic color scale. (c) Normalized radial displacements vs. the rescaled point force.

6. Boundary Instabilities and Wrinkling

An interesting feature observed in plant growth is the occurrence of self-similar wrinkles along free tissue boundaries, such as the edges of flowers and leaves [19, 20, 30–33]. The morphology is very similar to the shape of torn plastic sheets, and apparently, both phenomena are characterized by a plastic longitudinal metric profile $g_l(z) = 1/(1 + z/l)$, where $l > 0$ is a characteristic length scale and $z \geq 0$ is the coordinate perpendicular to the growing edge. Audoly and Boudaud [34] were able to show that the solution of the Föppl–von Kármán equations on the edge of a free rectangular sheet with such growth profiles consists of self-similar wrinkles governed by odd integral scaling factors. The Föppl–von Kármán equations are, however, geometrically limited as they don't allow reentrancy. The present growing Koiter shell model allows us to numerically solve the problem with its full geometric nonlinearity taken into account. Consider a flat rectangular sheet of thickness $h = 10^{-4}$, length $L = 4$, and width $W = 1$, initially lying in the xz plane. We clamp the long edge at $z = W$ and constrain the short edges to stay at $x = -L/2, L/2$, leaving them free to move in other directions. Plastic growth is imposed by setting the growth tensor to

$$\mathbf{F}_g = \mathrm{diag}(1 + g_l(z), 1, 1) \tag{12}$$

in Cartesian coordinates (x, y, z), and we choose a characteristic length $l = 40h$ for the growth field in this example. Fig. 4 shows the resulting equilibrated configuration after growth (or tearing), and a movie showing the equilibration is provided in the supplementary material. The self-similarity of the free edge at $z = 0$ is apparent, clearly resembling the wrinkling cascades observed in experiments [20, 30].

We have measured the fractal dimension of the grown edge depicted in Fig. 4(a) using the box counting method [35] and the self-similarity method [36]. In the former, the curve length L_{in} contained in a cubic box is determined as a function of the edge length L_{box} of the box. A fractal curve is expected to scale as $L_{\mathrm{in}} \sim L_{\mathrm{box}}^{D_f}$. Such scaling is indeed observed with fractal dimension $D_f = 1.15(1)$ (Fig. 5(a)). The scaling breaks down due to the influence of the clamped opposite edge, which introduces a global straightening effect when the box size is large ($L_{\mathrm{box}} \approx W$). The second method is more robust to global orientation and is thus better suited here. The length L_s of a piecewise linear path along the curve with segment size s is measured and expected to scale as $L_s \sim s^{1-D}$. We find a self-similarity dimension $D = 1.196(5)$ (Fig. 5(b)), which is very close

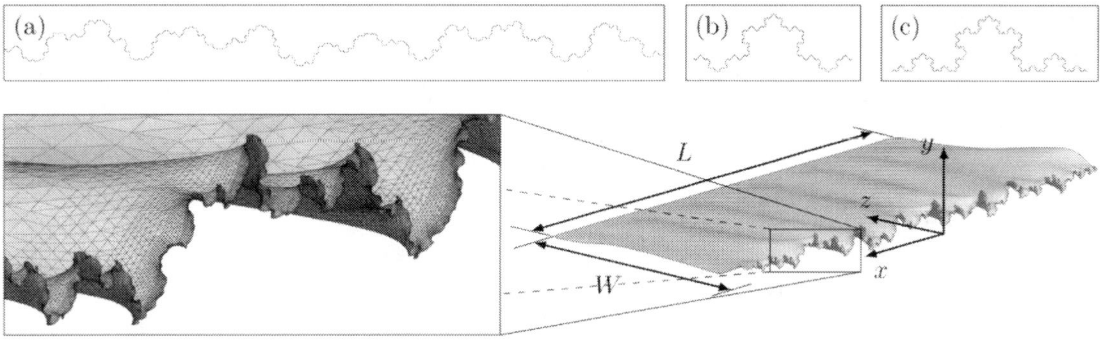

Figure 4. Self-similar sheet boundary after growing according to Eq. (12). (a) Projection of the grown edge onto the xy plane. (b) 60° Fibonacci word fractal. (c) Koch snowflake.

to the Hausdorff dimension of a 60-degree Fibonacci word fractal (Fig. 4(b)), $D_{\mathrm{H}} = 1.2083$ [37], and a bit lower than that of a triadic Koch curve (Fig. 4(c)), $D_{\mathrm{H}} = 1.2619$ [38].

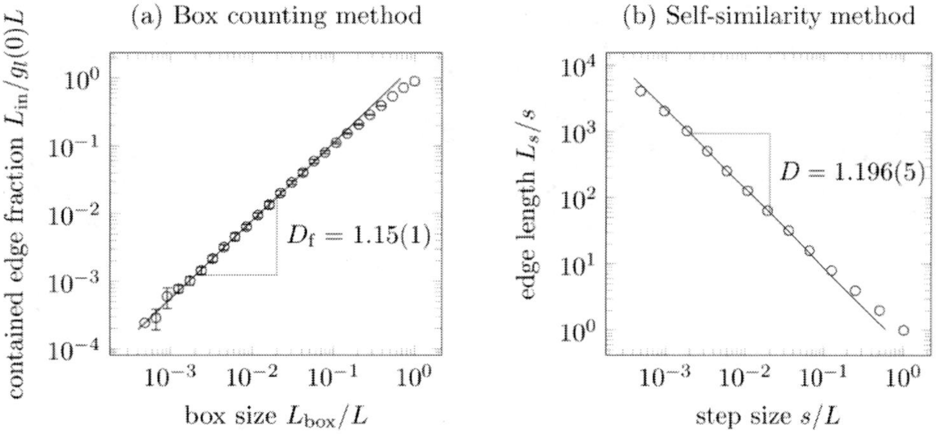

Figure 5. Fractal dimension of the edge at $z = 0$ of a thin sheet grown according to Eq. (12), measured with two standard methods.

The metric profile used above is not the only one yielding wrinkled edges. When it comes to wavy flowers like certain orchids for instance, single-wavelength undulations instead of self-similar edges are not uncommon. The feature causing wrinkle cascades is the presence of a non-constant geometric length scale defined by $l_{\mathrm{geo}}(z) = -g(z)/g'(z)$ [33]. The following families of growth fields will thus produce similar boundary instabilities:

$$g_{l,p}(z) \propto \left(1 + \frac{z}{p\,l}\right)^{-p}, \qquad (l > 0,\ p > 0,\ z \geq 0), \tag{13}$$

$$g_{l,p}(z) \propto \left(1 - \frac{z}{p\,l}\right)^{p}, \qquad (l > 0,\ p > 1,\ 0 \leq z \leq p\,l), \tag{14}$$

On the other hand, an exponential growth field

$$g_l(z) \propto \exp\left(-\frac{z}{l}\right), \qquad (l > 0,\ z \geq 0) \tag{15}$$

yields only a single wavelength [19] because $l_{\rm geo}(z) \equiv l$ in this case. On some flowers, these undulations may be forced to integral wavenumbers n by angular periodicity of a single petal. Let's hence significantly increase the characteristic length l and thickness h such that only a single wavelength λ prevails even for growth in the form of Eqs. (13,14), and let's consider a cylinder with height H and radius R instead of a flat plate. This change in geometry delivers dramatic consequences: A thin cylindrical sheet growing in circumferential direction according to $g(z)$, where z is the cylinder axis, only breaks its axisymmetry if growth leads to a circumference that changes faster than the sheet's metric can account for [31]. (Note that the excluded linear case $p = 1$ of Eq. (14) produces the excess cone [16,39] in the limit $R \to 0$, which doesn't wrinkle at the boundary because $g''_{l,1}(z) \equiv 0$.) A direct consequence of the Gauss–Bonnet theorem is that the axisymmetry is preserved as long as

$$\left| R \frac{\mathrm{d}g}{\mathrm{d}z}(z) \right| \leq 1, \quad 0 \leq z \leq H, \tag{16}$$

and broken otherwise. The origin for this instability is the (non)existence of *embeddings* of the surface: According to Gauss's Theorema Egregium, the creation or elimination of Gaussian curvature must be accompanied by in-plane stretching, which is traded for out-of-plane buckling if the sheet is sufficiently thin (see Fig. 6).

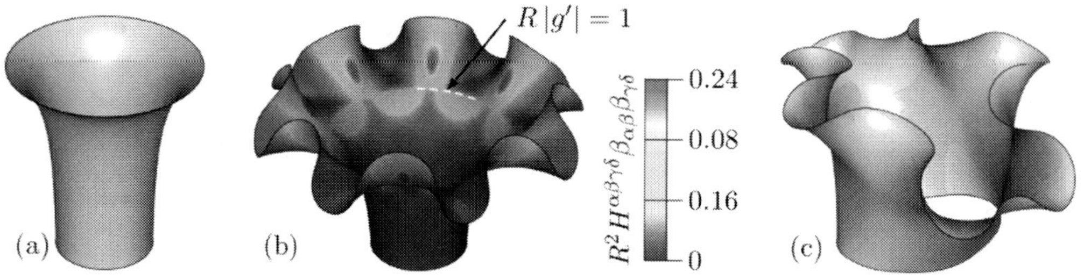

Figure 6. Boundary instability of a circumferentially growing cylinder. (a) As long as Ineq. (16) holds, axisymmetry is preserved. (b) Same growth field $g(z) \propto \exp(-z/l)$ as in (a), but larger prefactor. The axisymmetry is spontaneously changed to n-fold rotational symmetry C_n ($h/R = 0.02$, $l/R = 1$, $n = 8$). The rescaled bending energy density is shown in color, revealing the line where Ineq. (16) holds equally. (c) A relatively short cylinder (small H/l) with free boundaries also buckles away from the wrinkled edge, breaking C_n symmetry further to C_2.

How does the number n of boundary waves scale with l, h, R when axisymmetry is broken, and is it universal for all positive, monotonically decreasing and strictly convex growth profiles satisfying

$$\lim_{z \to 0} l_{\rm geo}(z) = l \;? \tag{17}$$

Since the preferred wavelength λ is a local feature independent of global geometry and topology (independent of R), we may use the ansatz [33]

$$\lambda \sim h^{\alpha}\, l_{\rm geo}^{1-\alpha}, \qquad \text{i.e.,} \qquad \frac{\lambda}{h} \sim \left(\frac{l_{\rm geo}}{h}\right)^{1-\alpha}. \tag{18}$$

On the other hand, geometry implies that

$$\lambda = \frac{2\pi R}{n}\left(1 + g(0)\right). \tag{19}$$

since $z = 0$ is where Ineq. (16) is violated first, given that $g' < 0$ and $g'' > 0$. After combining Eqs. (16–19), one thus finds a scaling for the number of wrinkles

$$n \sim \left(1 + \frac{R}{l}\right)\left(\frac{l}{h}\right)^{\alpha}. \tag{20}$$

Up to the first term, which accounts for the mean curvature of the cylinder, this coincides with the scaling law reported in [34] for the wrinkling hierarchies in initially flat sheets, where $\alpha = 2/5$ is found for the family $g_l(z) \propto 1/(1 + z/l)$. Our numerical data, which best fits Eq. (20) with $\alpha = 0.39(2)$, shows that the wrinkles studied here fall in the same category, see Fig. 7. Moreover, the data collapse of all employed growth profiles on a single line indicates that the scaling is universal in this respect. For this numerical study, we used

$$\mathbf{F}_{\mathrm{g}} = \mathrm{diag}\big(1, 1 + g(z), 1\big) \tag{21}$$

in cylindrical coordinates (r, φ, z) with various growth profiles proportional to Eqs. (13-15), and with slowly increasing proportionality prefactors. A movie showing such spontaneous wrinkling for different n is provided in the supplementary material.

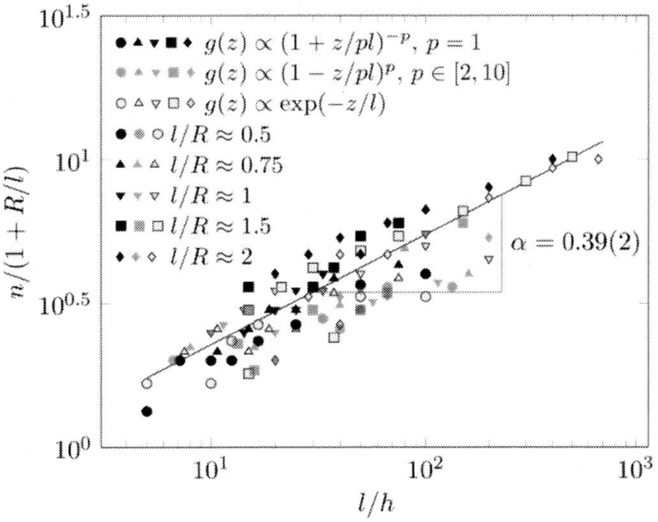

Figure 7. Scaling of the wrinkle number n along the edge of a circumferentially growing cylinder. The mode with the lowest energy is the integer n closest to the power law (20), but excited modes also randomly occur and are metastable (data points lying significantly above or below the straight line).

7. Confined Growth and Crumpling

Many numerical simulations of thin sheets getting folded and crumpled inside of shrinking hollow spheres have been carried out recently [4–6,40]. An important result is that the high bulk stiffness of crumpled sheets is due to a network of vertices and lines carrying large mean curvature and bending energy [1]. But what if instead of externally forced compression, the thin sheet intrinsically grows inside of a fixed spatial container? Are the processes that crumple a plant leaf or petal growing inside a bud the same as for a piece of foil crumpled by hand? Indeed they are in the elastic limit, as the following simulation demonstrates.

A thin circular sheet (radius R) is placed inside of a spherical cavity (radius $\overline{R}$). In the first setup, the container is shrunk, folding and crumpling the sheet into a ball of the size of the container. In the second setup, the container sustains its size while the sheet undergoes uniform isotropic growth, both in plane and in thickness. The only important parameter for this problem is the Föppl–von Kármán number $\gamma \sim (\overline{R}/h)^2 = 10^4$. Equivalent time scales are obtained by shrinking the sphere according to $\overline{R}(t) = R/(1+g(t))$, where $g(t) = \lambda t$ is the growth factor of the growing sheet. The growth rate λ is chosen small enough to keep inertial effects negligible. We add repulsive contact forces penalizing volumetric overlap between any two pieces of the sheet. Initially, both sheets buckle to form a developable cone with a single vertex (see Fig. 8, first column) that starts to nucleate at $\overline{R}/R \approx 0.53$. The emerging ridge network of focused mean curvature (third column) and bending energy (fourth column) is the same in both scenarios. Compared to similar measurements [5], the cross correlation $r = 0.89$ of the mean curvature ridge patterns is very high. The differences are only local and of the order of the mesh resolution. A movie showing the crumpling process is contained in the supplementary material.

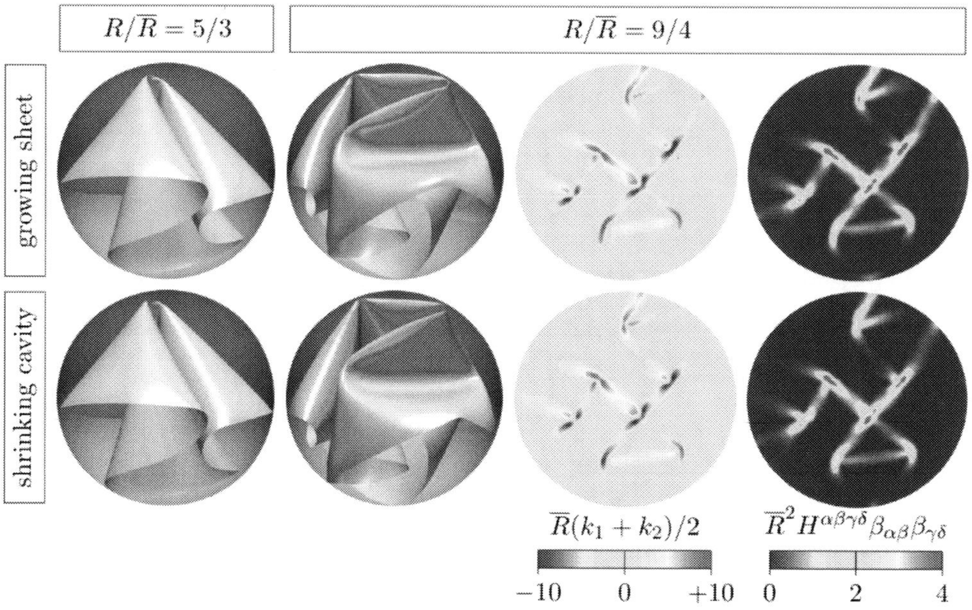

Figure 8. Comparison between shrinking confinement (bottom row) and growth in static confinement (top row), showing that the two processes are equivalent. k_1 and k_2 are the principal curvatures of the middle surface.

8. Conclusions

Simulating the plastic growth of thin sheets can be numerically demanding. The finite element method, being perhaps the best-suited technique for anisotropies, lacked an expedient C^1-continuous discretization, which is indispensible from a theoretical viewpoint, until subdivision surfaces were ported to it. We have extended the Kirchhoff–Love theory by arbitrary volumetric in-plane growth [25] and used the Loop subdivision surface paradigm to build a highly flexible and efficient numerical tool for the simulation of nonlinear thin sheet mechanics. Requiring no rotational variables, a thin sheet representation of this kind is remarkably simple to implement and superior to traditional approaches in terms of computational costs. A series of example simulations were carried out to demonstrate these strengths. In particular, we have quantified the self-similarity and scaling of wrinkles along plastically stretched free edges of thin sheets, and found that the problem of a growing confined sheet is equivalent to the narrowing confinement of a static sheet.

Acknowledgments

Financial support support from the ETH Research (ETHIIRA) Grant No. ETH-03 10-3 as well as from the European Research Council (ERC) Advanced Grant No. 319968-FlowCCS is gratefully acknowledged.

References

[1] Lobkovsky A, Gentges S, Li H, Morse D and Witten T A 1995 *Science* **270** 1482–1485
[2] Ben Amar M and Pomeau Y 1997 *Proc. R. Soc. A* **453** 729–755
[3] Blair D L and Kudrolli A 2005 *Phys. Rev. Lett.* **94**(16) 166107
[4] Vliegenthart G A and Gompper G 2006 *Nat. Mater.* **5**(3) 216–221
[5] Tallinen T, Åström J A and Timonen J 2008 *Phys. Rev. Lett.* **101**(10) 106101
[6] Tallinen T, Åström J A and Timonen J 2009 *Nat. Mater.* **8**(1) 25–28
[7] Vliegenthart G A and Gompper G 2011 *New J. Phys.* **13** 045020
[8] Otubushin A 1998 *Int. J. Impact Eng.* **21** 349–368
[9] Webb D C, Webster J and Kormi K 2001 *Int. J. Crashworthiness* **6** 399–424
[10] Mamalis A, Manolakos D, Ioannidis M, Kostazos P and Hassiotis G 2001 *Int. J. Crashworthiness* **6** 155–164
[11] Maia L G and de Oliveira P H I A 2005 *SAE Technical Paper*
[12] Love A E H 1888 *Phil. Trans. R. Soc. A* **179** 491–546
[13] Cirak F, Ortiz M and Schröder P 2000 *Int. J. Numer. Meth. Eng.* **47** 2039–2072
[14] Cirak F and Ortiz M 2001 *Int. J. Numer. Meth. Eng.* **51** 813–833
[15] Taber L A 1995 *Appl. Mech. Rev.* **48** 487–545
[16] Dervaux J and Ben Amar M 2008 *Phys. Rev. Lett.* **101**(6) 068101
[17] Hoger A 1986 *J. Elasticity* **16**(3) 303–324
[18] Skalak R, Zargaryan S, Jain R K, Netti P A and Hoger A 1996 *J. Math. Biol.* **34**(8) 889–914
[19] Marder M, Sharon E, Smith S and Roman B 2003 *Europhys. Lett.* **62** 498
[20] Marder M, Deega R D and Sharon E 2007 *Phys. Today* **60**(2) 33–38
[21] Koiter W T 1966 *Proc. Kon. Ned. Akad. Wetensch. B* **69** 1–54
[22] Ciarlet P 2000 *C. R. Acad. Sci. I* **331** 405–410
[23] Lee E H 1969 *ASME J. Appl. Mech.* **36** 1–6
[24] Rodriguez E K, Hoger A and McCulloch A D 1994 *J. Biomech.* **27** 455–467
[25] Vetter R, Stoop N, Jenni T, Wittel F K and Herrmann H J 2013 *Int. J. Numer. Meth. Eng.* **95** 791–810
[26] Hure J, Roman B and Bico J 2012 *Phys. Rev. Lett.* **109**(5) 054302
[27] King H, Schroll R D, Davidovitch B and Menon N 2012 *P. Natl. Acad. Sci. USA* **109** 9716–9720
[28] Negahban M, Goel A, Marchon P and Azizinamini A 2009 *Int. J. Comput. Meth. Eng. Sci. Mech.* **10** 430–449
[29] Wiśniewski K 2010 *Finite Rotation Shells: Basic Equations and Finite Elements for Reissner Kinematics* (Dordrecht: Springer Netherlands)
[30] Sharon E, Roman B, Marder M, Shin G S and Swinney H L 2002 *Nature* **419**(6907) 579
[31] Marder M 2002 ArXiv:cond-mat/0208232
[32] Audoly B 2004 *Europhys. News* **35** 145–148
[33] Sharon E, Roman B and Swinney H L 2007 *Phys. Rev. E* **75**(4) 046211
[34] Audoly B and Boudaoud A 2003 *Phys. Rev. Lett.* **91**(8) 086105
[35] Forrest S R and Witten T A 1979 *J. Phys. A: Math. Gen.* **12** L109
[36] Mandelbrot B 1967 *Science* **156** 636–638
[37] Monnerot-Dumaine A 2009 The Fibonacci Word Fractal `http://hal.archives-ouvertes.fr/hal-00367972`
[38] Mandelbrot B B 1982 *The fractal geometry of nature* (W. H. Freeman and Co.)
[39] Stoop N, Wittel F K, Ben Amar M, Müller M M and Herrmann H J 2010 *Phys. Rev. Lett.* **105**(6) 068101
[40] Kramer E M and Witten T A 1997 *Phys. Rev. Lett.* **78**(7) 1303–1306

Novel concepts for the systematic statistical analysis of phase transitions in finite systems

M Bachmann

Center for Simulational Physics, The University of Georgia, Athens, GA 30602, USA;
Instituto de Física, Universidade Federal de Mato Grosso, Cuiabá (MT), Brazil;
Departamento de Física, Universidade Federal de Minas Gerais, Belo Horizonte (MG), Brazil

E-mail: bachmann@smsyslab.org; http://www.smsyslab.org

Abstract. We review recent developments in the conceptual approach to a consistent systematic understanding of cooperative thermodynamic activity. The microcanonical statistical analysis is a powerful tool that is particularly useful for the investigation of analogs of phase transitions in finite systems, but it applies also in the extrapolation toward the thermodynamic limit. Whereas Maxwell construction in the coexistence region in the inverse caloric temperature space is a reasonable method for individual first-order-like transitions that are well-separated from other energetic regions of thermal activity, microcanonical inflection-point analysis is a systematic method that enables both the identification and classification of transitions of first and higher orders.

1. Introduction

Since the earliest observations that any physical system can exist in different thermodynamic phases, there has been a substantial interest in understanding the reasons for the qualitative changes that a macroscopic system is capable to experience while passing a transition point in the space of environmental parameters [1]. Conventional thermodynamics cannot explain the large-scale fluctuations and correlations that accompany a phase transition, but provides the basis for an adequate description that culminated in Ehrenfest's classification of phase transitions by locating discontinuities and divergences in thermodynamic variables (e.g., entropy, volume) and response quantities (specific heat, susceptibility, etc.), respectively. These quantities can be represented by distinct derivatives of appropriate thermodynamic potentials. For example, for a closed system of particles at constant temperature, pressure, and particle number, the thermodynamic potential is the Gibbs enthalpy $G(T,p,N)$. If we generally denote any thermodynamic potential that represents the thermodynamic system by $P(\mathbf{A})$, where $\mathbf{A}$ is the vector of independent external thermodynamic variables (such as T, p, and N in the example above), then a phase transition is observed at a point $\mathbf{A}_0$, if any of the derivatives of P with respect to any component of $\mathbf{A}$,

$$\left(\frac{\partial^{(n)}P}{\partial A_i^{(n)}}\right)_{\mathbf{A}_0} \tag{1}$$

is discontinuous or divergent at $\mathbf{A}_0$. The order of the transition is given by the lowest derivative n_0 that exhibits catastrophic behavior at the transition point. In most cases, only transitions of first ($n_0 = 1$) and second order ($n_0 = 2$) are distinguished. Each of the derivatives in Eq. (1)

can be understood as a response quantity. In our example, the response of G to a change of temperature is the entropy, $S(T,p,N) = -(\partial G/\partial T)_{p,N}$, and a discontinuity at a temperature T_0 signals a first-order phase transition. The second derivative of G with respect to T is related to the heat capacity at constant pressure, $C_p(T,p,N) = T(\partial S/\partial T)_{p,N} = -T(\partial^2 G/\partial T^2)_{p,N}$, which quantifies the response of the entropy to a change in temperature. A discontinuity or divergence of C_p at a critical temperature T_c indicates a second-order phase transition, provided the entropy is continuous at the transition point. Similar relations hold for responses to changes in the other natural thermodynamic variables of G.

However, Ehrenfest's theorem applies only in the thermodynamic limit, i.e., for very large systems, because thermodynamic quantities that describe the macrostate of finite systems do not exhibit any discontinuities or divergences. The interest in the understanding of thermodynamic activity in small or at least finite systems – with its extreme variant cooperativity – has rapidly grown recently. The sizes of nanotechnological applications and biological systems are microscopic or mesoscopic, i.e., surface effects affect the system's behavior to such an extent that the hypothetic extrapolation toward the thermodynamic limit is not helpful for the analysis of transitions in these systems. Therefore Ehrenfest's idea needs to be generalized.

The currently most promising approach is the so-called microcanonical analysis. Initially limited to first-order-like transitions by employing Maxwell's construction, the theory has recently been generalized in a systematic way that does not only allow for a unique identification of the transitions; it also enables the classification of the transitions similar to Ehrenfest's approach for large systems. This idea is based on the analysis of inflection points of the inverse microcanonical temperature.

In the following, we review the original idea of Maxwell's construction in the thermodynamic limit, its microcanonical version [1–10], and finally we discuss the basic features of the microcanonical inflection-point analysis [1, 11, 12].

2. Maxwell construction in thermodynamics

Transitions with phase coexistence, typically referred to as first-order phase transitions, are traditionally analyzed by means of Maxwell's construction. This method is necessary to remove the unphysical backbending branch that occurs for real gases described by the van der Waals equation in the pressure (p) versus volume (V) phase diagram for isotherms below the critical temperature $T_{\rm crit}$. Since $(\partial p/\partial V)_{T<T_{\rm crit}} > 0$ in this region, the compressibility is negative in the transition regime. The consequence would be a catastrophic instability of matter, although it is supposed to be in an equilibrium state. This is unphysical behavior and needs to be corrected. Figure 1(a) shows three van der Waals isotherms at temperatures above, at, and below the critical temperature.

It can be shown that in the thermodynamic limit the Gibbs free energies of the two coexisting phases (liquid, gas) must be identical for each macrostate in the transition region, $G_{\rm lg} = G_{\rm l} = G_{\rm g}$. The unphysical behavior of the van der Waals equation can only be avoided if the pressure in the transition region remains constant. The value of this pressure, $p_{\rm lg}$, is determined by the following consideration. Along the isotherm ($dT = 0$), $dG = Vdp$. Hence, if the Gibbs free energies in both phases would not be identical, the gradient ($V = (\partial G/\partial p)_T$) would spontaneously drive the system into one of the phases. This does not occur in a real gas-liquid mixture. The stable equilibrium liquid-gas phase coexistence can be observed. Additional work, as can be expressed by the change of free energy $dF = -pdV$, is necessary to change the macrostate (actually only V, because p and T are constant) within the transition region. Therefore, the total transition free energy released in the evaporation process is given by

$$F_{\rm lg} \equiv -\int_{V_{\rm l}}^{V_{\rm g}} p(V)dV = -p_{\rm lg}(V_{\rm g} - V_{\rm l}). \tag{2}$$

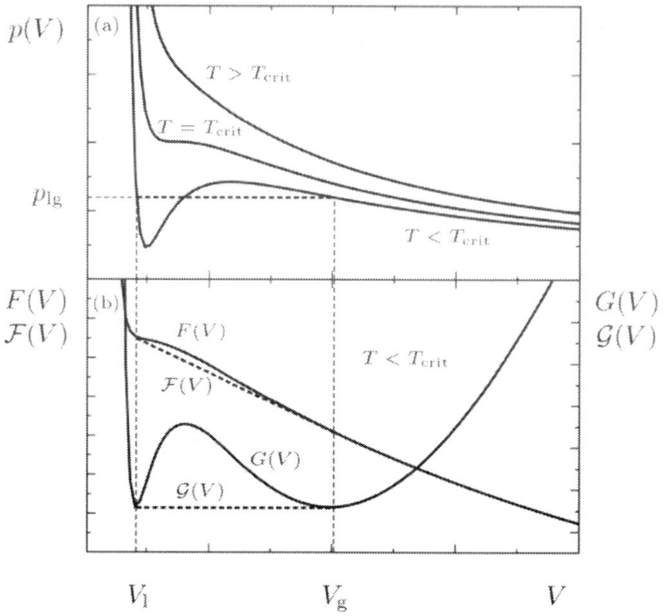

Figure 1. (a) Isotherms of the van der Waals gas $p(V)$ for temperatures T below, at, and above the critical temperature $T_{\rm crit}$, and Maxwell line at $p = p_{\rm lg}$. (b) Free energy $F(V)$, Gibbs construction for the free energy $\mathcal{F}(V)$ and Gibbs free energy $G(V)$ and its Gibbs hull $\mathcal{G}(V)$ at pressure $p = p_{\rm lg}$ for the isotherm in (a) at $T < T_{\rm crit}$.

where $V_{\rm l}$ and $V_{\rm g}$ denote the volumes at the respective phase boundaries, satisfying

$$\left(\frac{\partial F}{\partial V}\right)_{V_{\rm l}} = \left(\frac{\partial F}{\partial V}\right)_{V_{\rm g}} = -p_{\rm lg} = {\rm const} \tag{3}$$

in the interval $V \in [V_{\rm l}, V_{\rm g}]$. The area under the original curve $p(V)$ in this volume interval and the area bounded by $p_{\rm lg}$ are identical. The free energy as a function of the volume, $F(V)$, is shown in Fig. 1(b). Note the concavity $0 > (\partial^2 F/\partial V^2)_{T<T_{\rm crit}} = -(\partial p/\partial V)_{T<T_{\rm crit}} = 1/V\kappa$ of this function in the transition interval. Since neither V nor the compressibility κ can be negative, this behavior of F is unphysical.

The Maxwell construction for the subcritical isotherm corresponds to a Gibbs construction for the free energy: $\mathcal{F}(V) = F(V_l) - p_{\rm lg}(V - V_l)\ \forall V \in [V_{\rm l}, V_{\rm g}]$, i.e., the slope of this double tangent of $F(V)$ is $-p_{\rm lg}$. Since $\mathcal{F}(V) < F(V)\ \forall V \in [V_{\rm l}, V_{\rm g}]$, the macrostates $(p_{\rm lg}, V_{\rm lg} \in [V_{\rm l}, V_{\rm g}], T_{\rm lg} < T_{\rm crit})$ then indeed represent the stable and physical equilibrium states in the liquid-gas transition region. The hull of the Gibbs free energy, $\mathcal{G}(V) = \mathcal{F}(V) + p_{\rm lg}V$, is constant and minimal in the coexistence regime, as required. Thus, it replaces $G(V) = F(V) + p_{\rm lg}V$ as the appropriate thermodynamic potential for the van der Waals gas in the first-order transition region [see Fig. 1(b)]. As we have seen, the Maxwell construction is necessary for a physically consistent thermodynamic description of this model. It shall be noted that the problem lies in the equation of state. A statistical mechanics analysis of a microscopic model would not require to use Maxwell's construction method to "repair" unphysical behavior; the Maxwell line would establish itself in the thermodynamic limit.

Journal of Physics: Conference Series **487** (2014) 012013 doi:10.1088/1742-6596/487/1/012013

3. Microcanonical analysis by Maxwell construction for first-order-like transitions in finite systems

For the statistical analysis of finite systems that experience transitions that exhibit clear signatures of a first-order transition, a Maxwell construction can be rewarding, although it is not required for thermodynamic consistency.

The most fundamental statistical property and the fingerprint of any given physical system is the microcanonical entropy $S(E) = k_B \ln g(E)$, where $g(E)$ is the density of the states of the system. Typically, $S(E)$ is a concave function, except in a region of the energetic space, in which the number of energetic microstates is suppressed to such an extent that the entropy curve loses its concave monotony and becomes convex. Since in this depletion region the entropy has an energetic width $\Delta Q > 0$ and connects energetic spaces which are associated with different phases of reduced thermodynamic activity, the extended region in-between accommodates macrostates, in which both phases coexist. It is therefore common to interpret ΔQ as the latent heat and to associate the entropic suppression in this region with a first-order transition, in analogy to thermodynamic first-order phase transitions in the thermodynamic limit.

Figure 2(a) shows a typical example for an entropic curve with a convex region. The depletion zone is bounded by the limiting energies of the ordered phase, E_o, and the disordered phase, E_d. The width of the entropic depletion zone is $\Delta Q = E_d - E_o$. Also shown is the double-tangent (Gibbs hull), which corresponds to a Maxwell construction in the (inverse) temperature curve $\beta(E) = 1/T(E)$, as plotted in Figure 2(b). It enables the definition of a unique transition temperature β_{tr}. The unique identification of transition points is a notorious problem in conventional, typically canonical, analyses for finite systems [13]. The boundaries E_o and E_d are defined by the points, where the Gibbs hull

$$\mathcal{H}(E) = S(E_o) + \beta_{tr}(E - E_o) = S(E_d) + \beta_{tr}(E - E_d) \tag{4}$$

touches $S(E)$: $\mathcal{H}(E_{o,d}) = S(E_{o,d})$. The total entropy change during the transition $\Delta S_{tr} = S(E_d) - S(E_o) = \mathcal{H}(E_d) - \mathcal{H}(E_o)$ can then be expressed in two ways:

$$\Delta S_{tr} = \int_{E_o}^{E_d} dE\,\beta(E) = \beta_{tr}(E_d - E_o). \tag{5}$$

As Fig. 2(b) also shows, the β curve resembles a third-order polynomial with a "backbending" region. For that reason, the Maxwell line β_{tr} possesses a third intersection point with $\beta(E)$, denoted by E_{tr}. It can be considered as the separation point, where the entropic weight of each of the both phases in the coexistence region is identical. In this formalism, it defines the transition state. The total entropy change in the order-dominated coexistence regime is

$$\Delta S_o = \int_{E_o}^{E_{tr}} dE\,\beta(E) = S(E_{tr}) - S(E_o) = \beta_{tr}(E_{tr} - E_o) - s_o \tag{6}$$

and for the disordered part

$$\Delta S_d = \int_{E_{tr}}^{E_d} dE\,\beta(E) = S(E_d) - S(E_{tr}) = \beta_{tr}(E_d - E_{tr}) + s_d, \tag{7}$$

where the right-hand side expressions are simply obtained by adding the areas according to the definitions in Fig. 2(b). To resolve the meaning of s_o and s_d, and their relationship in

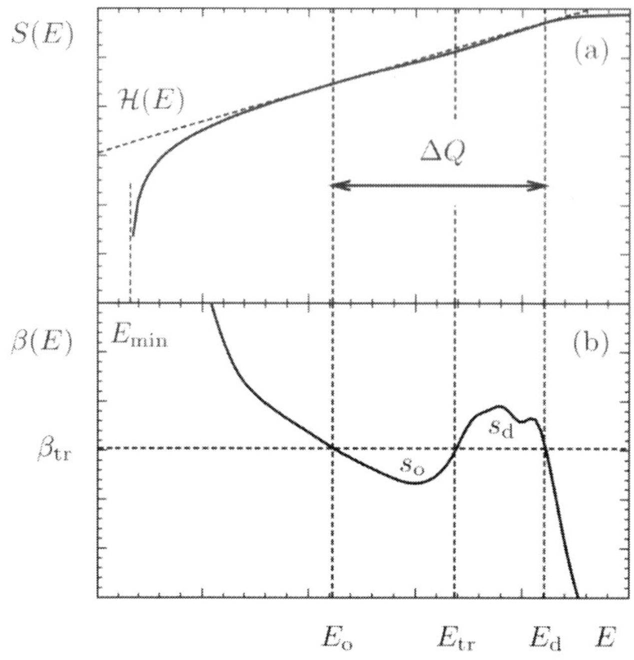

Figure 2. Microcanonical analysis by Maxwell construction: (a) Microcanonical entropy $S(E)$ and Gibbs construction $\mathcal{H}(E)$ as functions of energy E; (b) Inverse thermal energy $\beta(E)$ and Maxwell line at the first-order transition point $\beta_{\rm tr}$. The first-order-like transition behavior is characterized by the "backbending" effect between $E_{\rm o}$ and $E_{\rm d}$. The Maxwell line intersects $\beta(E)$ at energies $E_{\rm o}$, $E_{\rm tr}$, and $E_{\rm d}$. It is defined by the equality of the areas $s_{\rm o}$ and $s_{\rm d}$ which correspond to the surface entropy. These results were obtained by simulations of heteropolymer aggregation. Note the slight deformation of $\beta(E)$ right below $E_{\rm d}$ that indicates a subphase transition within the Maxwell regime, which cannot be considered separately in a microcanonical analysis based on Maxwell's construction.

correspondence with the Maxwell construction, we calculate the difference of the value of the Gibbs hull and the entropy,

$$\Delta S(E) = \mathcal{H}(E) - S(E), \tag{8}$$

at the separation point $E_{\rm tr}$:

$$\Delta S(E_{\rm tr}) = S(E_{\rm o}) + \beta_{\rm tr}(E_{\rm tr} - E_{\rm o}) - S(E_{\rm tr}) = S(E_{\rm d}) + \beta_{\rm tr}(E_{\rm tr} - E_{\rm d}) - S(E_{\rm tr}), \tag{9}$$

where we made use of the two parametrizations of the Gibbs tangent given in Eq. (4). Since $\Delta S(E_{\rm tr}) \neq 0$ is characteristic for first-order-like transitions in finite systems, where surface effects are not negligible, it is often called surface entropy, $\Delta S_{\rm surf} \equiv \Delta S(E_{\rm tr})$. Compared to the bulk of topological dimension D, the conformational entropic freedom at the surface of such systems is smaller, because of the reduced dimension $D-1$. This has the effect that particles prefer to be part of the interior volume and not of the surface layer of the system. This results in shapes with minimal surfaces (e.g., spherical or globular conformations in an isotropic

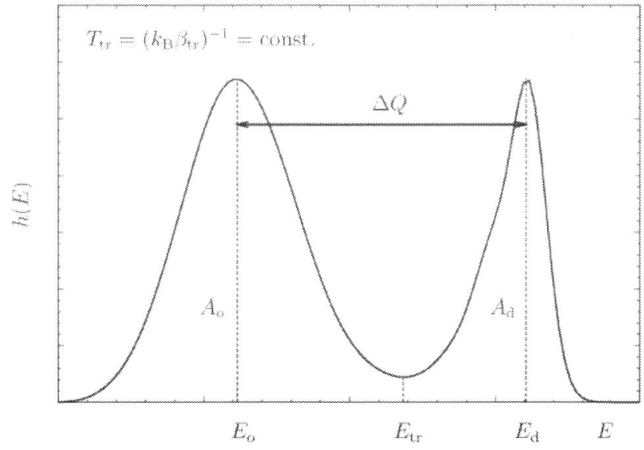

Figure 3. Canonical energy histogram at $T_{\rm tr} = (k_{\rm B}\beta_{\rm tr})^{-1}$ of the same transition as identified in Fig. 2. Note that it is common to define $\beta \equiv 1/k_{\rm B}T$ in the canonical analysis, whereas $\beta(E) \equiv 1/T(E)$ in the microcanonical theory.

medium). The "surface entropy" $\Delta S_{\rm surf}$ is actually the amount of entropy suppression by the formation of surfaces. If the effective attractive interactions between the particles have a finite range, surface effects decrease with increasing system size (number of particles N) and such does $\Delta S_{\rm surf}$. For such systems, in the thermodynamic limit, $\lim_{N\to\infty} \Delta S_{\rm surf} = 0$. Thus, in this limit, the entropy cannot be convex and thermodynamic first-order phase transitions do not exhibit this "backbending" feature that we discuss here in this section.

We eliminate $S(E_{\rm tr})$ in Eq. (9) by employing Eqs. (6) and (7) and eventually find that

$$\Delta S_{\rm surf} = s_{\rm o} = s_{\rm d}. \tag{10}$$

Thus, the Gibbs construction $\mathcal{H}(E)$ that corresponds to Maxwell's line at $\beta_{\rm tr}$, automatically leads to identical areas $s_{\rm o}$ and $s_{\rm d}$ and requires these areas to be identical to the surface entropy $\Delta S_{\rm surf}$. This is a remarkable result, as it shows that the surface entropy corresponds to the maximum deviation of the Gibbs tangent from the original $S(E)$ curve, located at $E_{\rm tr}$. The transition point between the ordered and the disordered phase is well-defined by $\beta_{\rm tr}$. Thus, the microcanonical analysis by Maxwell construction does not possess any ambiguity in the definition of transition points for finite systems, in contrast to the canonical analysis of peak positions and "shoulders" of fluctuations of typically not well-defined order parameters [1, 13].

The energy is considered here as an order parameter. It is not very specific (we do not learn much about the physical properties of the transition), but it allows us to clearly separate the phases. This is particularly apparent when falling back to the canonical interpretation of these features. Figure 3 shows the canonical histogram (unnormalized probability density distribution) $h(E) = g(E)e^{-E/k_{\rm B}T_{\rm can}}$ at the transition temperature $T_{\rm can} = T_{\rm tr}$ identified by Maxwell construction. This distribution has a characteristic bimodal form, with peaks at energies $E_{\rm o}$ and $E_{\rm d}$, separated by the latent heat ΔQ. The minimum between the peaks is located at the separation point $E_{\rm tr}$. Note that this approach defines the transition point at $T_{\rm tr}$, where $h(E_{\rm o}) = h(E_{\rm d})$. This is, as we have seen before, a direct consequence of the microcanonical analysis. The areas under the peaks in the ordered and in the coexisting disordered phase, $A_{\rm o}$ and $A_{\rm d}$, respectively, are not necessarily identical.

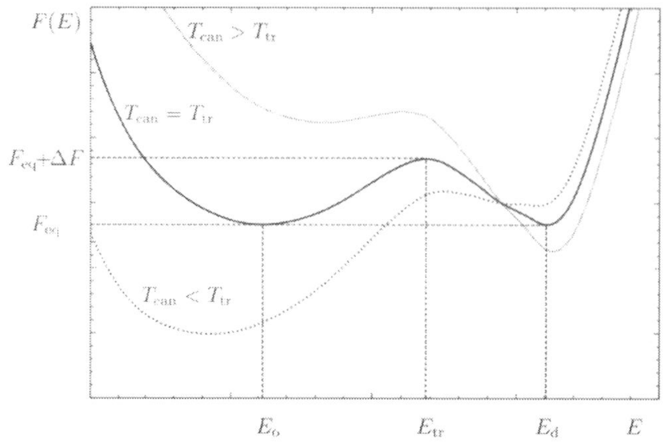

Figure 4. Free-energy profiles $F(E)$ for temperatures below, at, and above the transition point $T_{\rm can} = T_{\rm tr}$.

The histogram $h(E)$ represents the restricted partition sum of the system in energetic space,

$$h(E) \equiv Z^{\rm res}_{\rm can}(E) = \int dE'\delta(E-E')g(E')e^{-E'/k_{\rm B}T_{\rm can}}, \tag{11}$$

and $T_{\rm can}$ is a constant parameter representing the canonical temperature. Then, the free-energy landscape in energy space is parametrized by

$$F(E) = -k_{\rm B}T_{\rm can}\ln h(E) = E - T_{\rm can}S(E). \tag{12}$$

Thermodynamically, only global minima in the free-energy landscape represent stable equilibrium states. The extremum condition $(dF/dE)_{E=E_{\rm ext}} = 0$ thus yields

$$T(E_{\rm ext}) = T_{\rm can} \tag{13}$$

and connects the microcanonical and canonical pictures. Extrema in $h(E)$ at $T_{\rm can}$ are thus mirrored by extremal points in the $F(E)$ landscape, too. Because of the Maxwell construction, the transition point is defined by $T(E) = T_{\rm tr}$. This is only satisfied at energies $E_{\rm o}$, $E_{\rm tr}$, and $E_{\rm d}$, which, according to the condition (13) are also the locations of extremal points in the free-energy landscape at the transition point. In this interpretation, $F(E_{\rm o}) = F_{\rm eq}$ and $F(E_{\rm d}) = F_{\rm eq}$ represent the degenerate macrostates of the coexisting phases, and $F(E_{\rm tr}) = F_{\rm eq}+\Delta F$ is the free energy of the transition barrier. The height of the barrier, ΔF can be interpreted kinetically. Figure 4 shows free-energy profiles for temperatures below the transition point $T_{\rm can} < T_{\rm tr}$, where the system is in the ordered phase, at the microcanonically defined transition point $T_{\rm can} = T_{\rm tr}$, where the ordered and the disordered phases coexist, and above the transition temperature ($T_{\rm can} > T_{\rm tr}$). In the latter case, the free energy minimum lies beyond $E_{\rm d}$, i.e., the system resides in the disordered phase.

Note that if one replaces in Eq. (12) the entropy $S(E)$ by its Gibbs hull in the transition regime,

$$F(E) = E - T_{\rm tr}\mathcal{H}(E) = E_{\rm o} - T_{\rm tr}S(E_{\rm o}) = E_{\rm d} - T_{\rm tr}S(E_{\rm d}) \tag{14}$$

$$= F_{\rm eq} = {\rm const.} \quad \forall E \in [E_{\rm o}, E_{\rm d}]. \tag{15}$$

This is how the Maxwell construction affects the free-energy behavior: Each macrostate in the coexistence regime possesses the same free energy. This resembles the behavior of the Gibbs free energy in the liquid-gas coexistence regime of the van der Waals gas [see Fig. 1(b)]. Remember that in the latter case, the Maxwell construction was a physical necessity in the thermodynamic limit. In the microcanonical analysis for finite systems, we only made use of it to determine the transition point. We can therefore argue that the transition barrier $\Delta F > 0$ for the finite system is physical reality, but it disappears in the thermodynamic limit and the $F(E)$ curve converges to the Maxwell line. It is worth noting at this point, that the kinetics of processes such as protein folding is strongly affected by the existence of a transition barrier. Nonetheless, it is only a finite-size effect because of its surface-entropic origin.

The general disadvantage of the microcanonical analysis based on Gibbs/Maxwell construction is that it is only applicable to transitions with a clear, single signature, the "backbending effect". Often, this is actually not the case. First-order transitions typically exhibit hierarchies of individual subphase transitions [1, 14]. A first indication can already be noticed in the example shown in Fig. 2(b), where an additional mini-backbending effect [6] can be observed between $E_{\rm tr}$ and $E_{\rm d}$. In a systematic microcanonical analysis, such signals shall not be ignored. However, since this transition lies within the Maxwell coexistence regime, it cannot be resolved by this kind of analysis. Transitions without backbending feature, which we will later define as transitions of higher than first order, cannot be analyzed by Maxwell construction as well. In order to enable a thorough and systematic microcanonical analysis that includes hierarchies in first-order transitions and the identification of higher-order transitions, a revised theory is needed. A straightforward approach will be discussed in the following.

4. Systematic approach to the unique identification and classification of transitions in finite systems

As shown, the Maxwell construction can successfully be employed for the analysis of strong first-order transitions. Geometrically, the backbending region in the transition regime of the energetic temperature curve in Fig. 2 is replaced by an entirely flat segment. We also know that the backbending effect is due to surface effects that become negligible for very large systems, where volume properties are dominant. Thus, this flattening of the $\beta(E)$ curve is a physical property of this transition. It means that, in the transition region, the (volume part of the) system is not very sensitive to energetic changes. Macrostate changes require more energy (latent heat) while temperature changes decrease. The kinetics of macrostate changes is affected by the transition barrier. However, the Maxwell construction only applies to individual transitions of first order.

To include other transition types in the microcanonical analysis as well, it is attractive to extend the of idea of a "flat" energetic region by replacing the Maxwell construction by a more general principle, the *principle of least sensitivity*, which dates back to the early days of classical mechanics. This is a weaker condition, but it allows us to investigate first- and higher-order transitions by a microcanonical analysis more systematically and in much more detail [1, 11].

This approach takes into account that all qualitative changes in the interplay of entropy and energy as signaled by alterations in the curvature of the microcanonical entropy $S(E)$, are indicators of cooperative behavior in the system. Since the complete phase behavior is already encoded in $S(E)$, the inverse microcanonical temperature $\beta(E)$ is considered to be the only unique parameter to identify transition points. The monotony of $\beta(E)$, as expressed by its derivative with respect to energy,

$$\gamma(E) = d\beta(E)/dE = d^2S/dE^2. \tag{16}$$

allows for the introduction of a systematic classification scheme of transitions in finite systems [11], but it can also be used for scaling analyses towards the thermodynamic limit. The function $\gamma(E)$ describes the variation of the inverse temperature with respect to energy

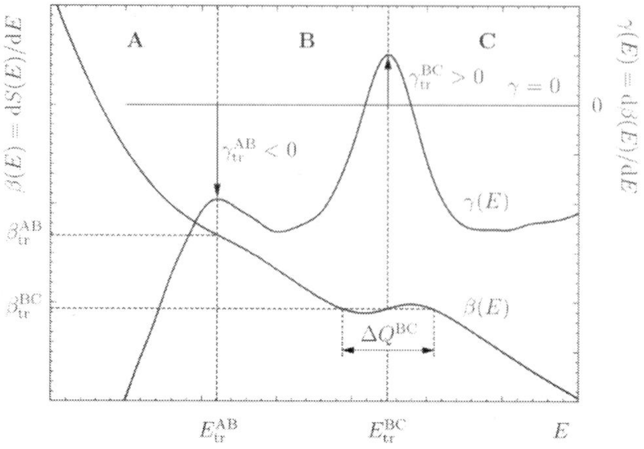

Figure 5. Inflection-point analysis of the inverse temperature $\beta(E)$ and its derivative $\gamma(E)$ as functions of energy E for a system exhibiting a first- and a second-order transition. The maxima of $\gamma(E)$ indicate transitions between the phases A and B at E_{tr}^{AB} and B and C at E_{tr}^{BC}. The associated points $\beta(E_{tr}^{AB}) = \beta_{tr}^{AB}$ and $\beta(E_{tr}^{BC}) = \beta_{tr}^{BC}$ define the transition temperatures $T_{tr}^{AB} = (\beta_{tr}^{AB})^{-1}$ and $T_{tr}^{BC} = (\beta_{tr}^{BC})^{-1}$. According to the classification scheme of inflection-point analysis, the transition between A and B is of second order, since the slope of the inflection point is negative. On the other hand, B$\leftrightarrow$C is a first-order transition as the respective slope at $\beta(E_{tr}^{BC})$ is positive. The non-monotonic "backbending" is a characteristic signal of phase coexistence. The latent heat ΔQ^{BC} is defined as the energetic width of this transition region.

at a given energy value E. As such it is related with the microcanonical heat capacity via $C_V(E) = [dT(E)/dE]^{-1} = -\beta^2(E)/\gamma(E)$.

In this scheme, a transition between phases is defined to be of *first order* if the slope of the corresponding inflection point of $\beta(E)$ at $E = E_{tr}$ is positive,

$$\gamma_{tr} = \gamma(E_{tr}) > 0 : \text{ first-order transition.} \tag{17}$$

Only in this case is the temperature curve non-monotonic and there is no unique mapping between β and E. As it has been discussed in the context of the Maxwell construction, this is the regime where both phases coexist. The overall energetic width of the undercooling, backbending, and overheating regions is identical to the latent heat. Thus, for a first-order transition, $\Delta Q > 0$. Note that the inflection point is not necessarily identical with the separation point obtained by Maxwell construction. A noticeable deviation between these energetic separation points is a strong indication that a subphase transition in one of the coexisting phases occurs. Such a transition cannot be detected by Maxwell construction only.

In the case that the inflection point has a negative slope, the phases cannot coexist and the latent heat is zero, $\Delta Q = 0$. In complete analogy to phase transitions in the thermodynamic limit, such transitions are classified as of *second order*:

$$\gamma_{tr} = \gamma(E_{tr}) < 0 : \text{ second-order transition.} \tag{18}$$

Since the inflection points of $\beta(E)$ correspond to maxima in $\gamma(E)$, it is therefore sufficient to analyze the peak structure of $\gamma(E)$ in order to identify the transition energies and temperatures. The sign of the peak values classifies the transition. This very simple and general classification scheme applies to all physical systems.

Figure 5 illustrates the procedure for the identification of the transitions by means of inflection-point analysis. Plotted are the inverse temperature $\beta(E)$ and its energetic derivative $\gamma(E)$. There are two regions, where the weak-sensitivity condition applies to $\beta(E)$. One is located around the inflection point at $E_{\rm tr}^{\rm AB}$ and the other is the backbending regime surrounding the central inflection point at $E_{\rm tr}^{\rm BC}$. The latter exhibits the already well-described features of a first-order transition: $\gamma_{\rm tr}^{\rm BC}$ is positive and the intersection points of the inverse transition temperature $\beta_{\rm tr}^{\rm BC}$ with the $\beta(E)$ curve define the coexistence region. The width is interpreted as the latent heat, which is obviously nonzero: $\Delta Q^{\rm BC} > 0$. The behavior is qualitatively different at $E_{\rm tr}^{\rm AB}$, where $\gamma_{\rm tr}^{\rm AB} < 0$. There is no phase coexistence so that the latent heat is zero. The A$\leftrightarrow$B transition is consequently classified as a second-order transition. The transition point is given by the inverse temperature $\beta_{\rm tr}^{\rm AB}$.

In cases, where it is difficult to find first- or second-order traces of "hidden" finite-size effects in $\beta(E)$, the weak-sensitivity criterion can be extended by allowing higher-than-second-order transitions. This can be done by analyzing the inflection points of $\gamma(E)$, or if necessary, of even higher derivatives of $\beta(E)$. Therefore, our hierarchical analysis is an analog to the previously discussed Ehrenfest scheme in thermodynamics.

There is another issue worth being remarked. The consistency of the hierarchy $S \to \beta = dS/dE \to \gamma = d^2/dE^2 \to \ldots$ tells us that, in principle, the *inverse* temperature β is more fundamental than the temperature T. This is not a trivial statement. Second-order transition points cannot easily be identified as inflection points of $T(E)$. The introduction of the temperature T (or, more precisely, $k_{\rm B}T$) as a "measure" for thermal energy had merely occurred as a historical accident, but there were no substantial physical reasons to favor it. Microcanonical analysis based on the close relationship between entropy and *inverse* temperature suggests to consider β as the more appropriate thermodynamic variable.

5. Conclusion

As a general principle, one can conclude that cooperative behavior is encoded in the curvature of an appropriate "thermodynamic potential". In the thermodynamic limit, one might want to choose to analyze phase transitions by means of canonical statistical analysis on the basis of the free energy. In the microcanonical analysis of small systems, it is the entropy as a function of energy that is considered to be the appropriate quantity. In the hierarchical inflection-point analysis, the variational principle standing behind the weak-sensitivity criterion tells us that first-order transitions are visible in the first derivative of $S(E)$ (backbending effect or Maxwell construction), second-order transitions are characterized by specific inflection points of $S(E)$, i.e., second-order derivatives are needed for their characterization, and higher-order transitions require higher-order derivatives to be identified. Inflection-point analysis is a simple but effective tool for a thorough and unique identification and classification of transitions in any physical system on all scales, from very small to very large.

Acknowledgement

Support by the NSF under Grant No. DMR-1207437 and by CNPq (National Council for Scientific and Technological Development, Brazil) under Grant No. 402091/2012-4 is gratefully acknowledged.

References

[1] Bachmann M 2014 *Thermodynamics and Statistical Mechanics of Macromolecular Systems* (Cambridge: Cambridge University Press), in press
[2] Thirring W 1970 *Z. Phys.* **235** 339
[3] Janke W 1998 *Nucl. Phys. B* **63A-C** 631
[4] Kastner M, Promberger M, and Hüller A 2000 *J. Stat. Phys.* **99** 1251
[5] Gross D H E 2001 *Microcanonical Thermodynamics* (Singapore: World Scientific)

[6] Junghans C, Bachmann M, and Janke W 2006 *Phys. Rev. Lett.* **97**, 218103
[7] Hernández-Rojas J and Gomez-Llorente J M 2008 *Phys. Rev. Lett.* **100**, 258104
[8] Junghans C, Bachmann M, and Janke W 2008 *J. Chem. Phys.* **128**, 085103
[9] Taylor M P, Paul W, and Binder K 2009 *Phys. Rev. E* **79** 050801(R)
[10] Taylor M P, Paul W, and Binder K 2009 *J. Chem. Phys.* **131** 114907
[11] Schnabel S, Seaton D T, Landau D P, and Bachmann M 2011, *Phys. Rev. E* **84**, 011127
[12] Gross J, Neuhaus T, Vogel T, and Bachmann M 2013, *J. Chem. Phys.* **138** 074905
[13] Bachmann M and Janke W 2004 *J. Chem. Phys.* **120** 6779
[14] Junghans C, Janke W, and Bachmann M 2011 *Comp. Phys. Commun.* **182** 1937

Journal of Physics: Conference Series **487** (2014) 012014 doi:10.1088/1742-6596/487/1/012014

Graphical Visualization on Computational Simulation Using Shared Memory

A. B. Lima[1] **and Eberth Correa**[2]

[1]Instituto de Ciências Tecnológicas e Exatas, Universidade Federal do Triângulo Míneiro - UFTM, 38064-200, Uberaba-MG, Brazil.

[2]Faculdade UnB Gama, Universidade de Brasília - UnB, 72444-240, Gama-DF, Brazil.

Abstract. The Shared Memory technique is a powerful tool for parallelizing computer codes. In particular it can be used to visualize the results "on the fly" without stop running the simulation. In this presentation we discuss and show how to use the technique conjugated with a visualization code using openGL.

1. Introduction

The visualization is an interesting tool to help us to understand some aspects of simulations [1–4]. In many cases we want to see the result of the simulations on the fly, so that we can decide whether the written code is correct or whether the set of chosen parameters was adequate in the actual simulation. In practice, visualization on the fly requires more computational effort than the simulation itself. As an alternative, one can dedicate some computational effort for the visualization in a separated program in such way that we can turn on or turn off the visualization when we want to. This technique is considerably advantageous, for the processing time is spent only at the visualization process.

The parallelization of the code is the more efficient way to proceed in this case. There are many ways to parallelize the code, we want to emphasize two of them: Message Passing Interface (MPI) and Shared Memory [5–8].

In the MPI technique, a copy of the code is put at each memory of each cpu and messages are exchanged between cpus (using a network connection) to set what each cpu must to do in the code. At the beginning, MPI and its library implementations (OpenMPI, MPICH, etc.) were the most useful way to implement parallelization in computer programs.

In the Shared Memory, different parts of the program use the same memory area to exchange data. Although unlike in concept, the Shared Memory parallelization has become useful with the recent development of more powerful multi-core processors found in either current personal computers or workstations, in which some of them can have up to 64 cores. It is worth emphasizing that the algorithms and programming implementation using Shared Memory is easier than MPI. A schematic view of MPI and Shared Memory is displayed in Fig. (1).

Shared Memory is one of the simplest method of interprocess communication (IPC) and allows two or more processes to exchange data accessing the same area in memory. In addition, these communications use the bus of the computer chipset being the fastest way to parallelize tasks and to avoid copying data unnecessarily.

(a) (b)

Figure 1. The (a)MPI and (b)Shared Memory schemes.The data exchange in Shared Memory is fastest that MPI.

As an example we present in the next section a standard algorithm to implement the Shared Memory in a computational simulation using Molecular Dynamics. In what follows, all the applications displayed use graphical visualizations in *openGL* for a molecular dynamics simulation of a magnetic liquid.

2. Shared Memory

To use the Shared Memory we need to allocate a memory segment. The function in C or C++ to allocate memory is called shmget ("SHared Memory GET"). Its first parameter is an integer that identifies which segment must be created. All processes can access the same memory segment by specifying this key. The second parameter specifies the number of bytes in memory segment. The third parameter is related to the flag values that specify several options to the shmget function. In the example below we show a C code fragment that creates a memory segment.

```
#include <sys/ipc.h>
#include <sys/shm.h>

#define SHMSZ 100 //size of memory segment (100 bytes)

int main(int argc, char* argv[])
{

int key,shmid;

key  = 11; //identifies the memory segment

// shmget: the flags IPC_CREAT | 0666 check if the segment
//exist and set the segment as a read and write. See
// more flags in shmget help.
if ((shmid = shmget(key, SHMSZ, IPC_CREAT | 0666 )) < 0) {
                perror(''shmget'');
                return 1;
        }
```

```
return 0;
}
```

Next we need to make the Shared Memory segment available. We must use shmat,("SHared Memory ATach"). This function uses as a first argument the identifier shmid returned by the shmget.The second argument is a pointer that specifies where inside our processes address we want to map the Shared Memory. It is easier to let the operational system decide what to do. This can be done only specifying it as NULL. The third argument is a flag. We show in the following the code to be used to attach the memory segment.

```
#include <sys/ipc.h>
#include <sys/shm.h>

int main(int argc, char* argv[])
{

double *shm;


if ((shm=(double *)shmat(shmid, NULL, 0)) == (double *) -1){
                  perror(''shmat'');
                  return 1;
return 0;
}
```

Now you can use the memory segment shm like a double array.

```
int main(int argc, char* argv[])
{

...

for(int i=0; i<number_of_double;i++){
shm[i]=0.0;
}

...

return 0;
}
```

To detach the Shared Memory segment we use the function shmdt.

```
int main(int argc, char* argv[])
{

...

shmdt(shm);
...
```

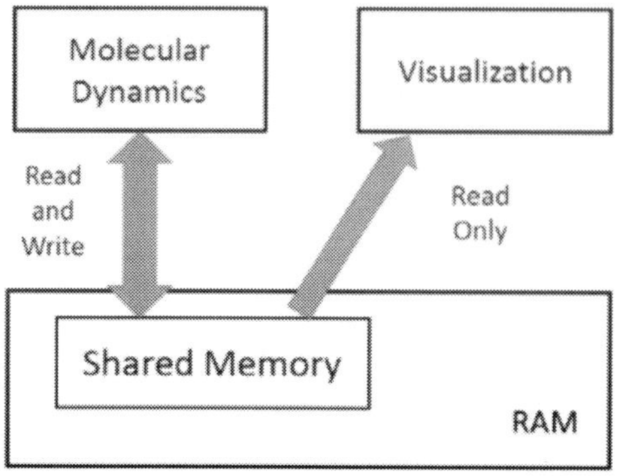

Figure 2. The Molecular Dynamics program can read and write in Shared Memory, but Visualization program just can read.

```
return 0;
}
```

To remove a Shared Memory segment we need to use the function shmctl ("SHared Memory Control"). This function uses the id of a segment as a first parameter, a flag IPC_RMID as a second argument and a NULL pointer as a third.

```
int main(int argc, char* argv[])
{

...

shmctl(shmid,IPC_RMID,NULL);
...

return 0;
}
```

For pedagogical purposes, we employ this technique to visualize a Molecular Dynamics simulation (MD). To make a real time simulation we need to put all these functions with "create", "attach", "detach" and "destroy" Shared Memory segments into the simulation code. Typically we need to use these functions to create 3 arrays of double precision to store x, y and z, the spatial coordinates, and 3 arrays of double precision to store Vx, Vy and Vz for the velocities. Fig. (2) shows how the MD simulation and Visualization works with a Shared Memory.

In the visualization program we use the same code as before, but in the third parameter of shmget we use a flag "read only" to prevent the visualization program to change any value in the arrays of the positions or the velocities.

```
#include <sys/ipc.h>
```

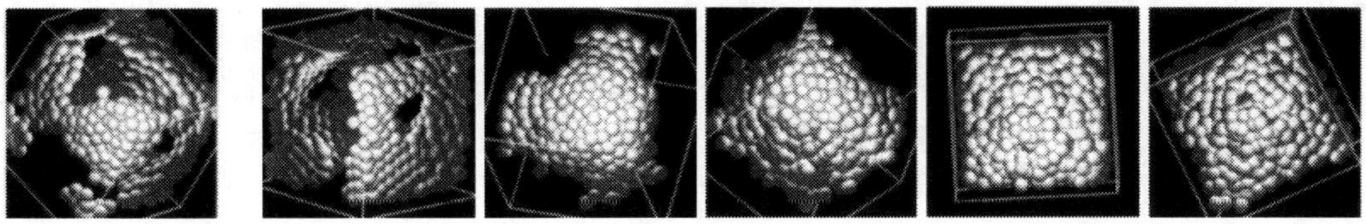

(a) (b) (c) (d) (e) (f)

Figure 3. (Color online) The visualization configurations for several temperatures, e.g. $T = 0.01, 0.2, 0.4, 0.6, 0.8, 1.0$.

```
#include <sys/shm.h>

#define SHMSZ //size of memory segment

int main(int argc, char* argv[])
{

int key,shmid;
double *shm;

key = 11; //identifies the memory segment

// shmget: the flags IPC_CREAT | 0444 check if the segment
//exist and set the segment as a read only. See
// more flags in shmget help.
if ((shmid = shmget(key, SHMSZ, IPC_CREAT | 0444 )) < 0) {
                perror(``shmget'');
                return 1;
        }

//attach the segment shm in program
if ((shm=(double *)shmat(shmid, NULL, 0)) == (double *) -1){
                perror(``shmat'');
                return 1;
//here we can read the value of shm
for(int i=0; i<number_of_double;i++){
printf(``%d/n'',shm[i]);
}

...

return 0;
}
```

To visualize the MD simulation we use the *freeglut* library in *openGL* [10]. In Fig. (3) we display some configurations after the equilibration of the system for temperatures varying from $T = 0.01$ to $T = 1.0$. Here we use reduced units [9] with the temperature given in units of

$\varepsilon/k_B \approx 120\,K$ and distances in units of σ, where ε and σ are the Lennard-Jones parameters and k_B the Boltzmann constant.

In a system with a considerable number of particles it is important to optimize the computational effort during the calculations. In a multi-core architecture the operational system does the tough task of separating the jobs between the cores. This does not forbid us to use other parallel schemes like the multithread technique, but this is out of scope of this work. With the help of the Shared Memory scheme we are able to switch on all the visualizations we want to see, separated from the computation of the MD simulations. As mentioned before we need only to enumerate the Shared Memory for each calculation in both the MD simulation and the *openGL* visualization programs without any interference between each other.

To illustrate the application we plot some radial distribution functions for the visualised configurations in Fig (4). The formation of the clusters can be seen as the temperature is decreased. However, in Fig. (3) we can see not only the clusters but also the structure formed in real time. The advantage of visualizing the simulations becomes more apparent in systems with magnetic properties like magnetic fluids [9]. Some characteristics are intrinsic for these systems like domain walls and vortex patterns in which the visualization is better suited.

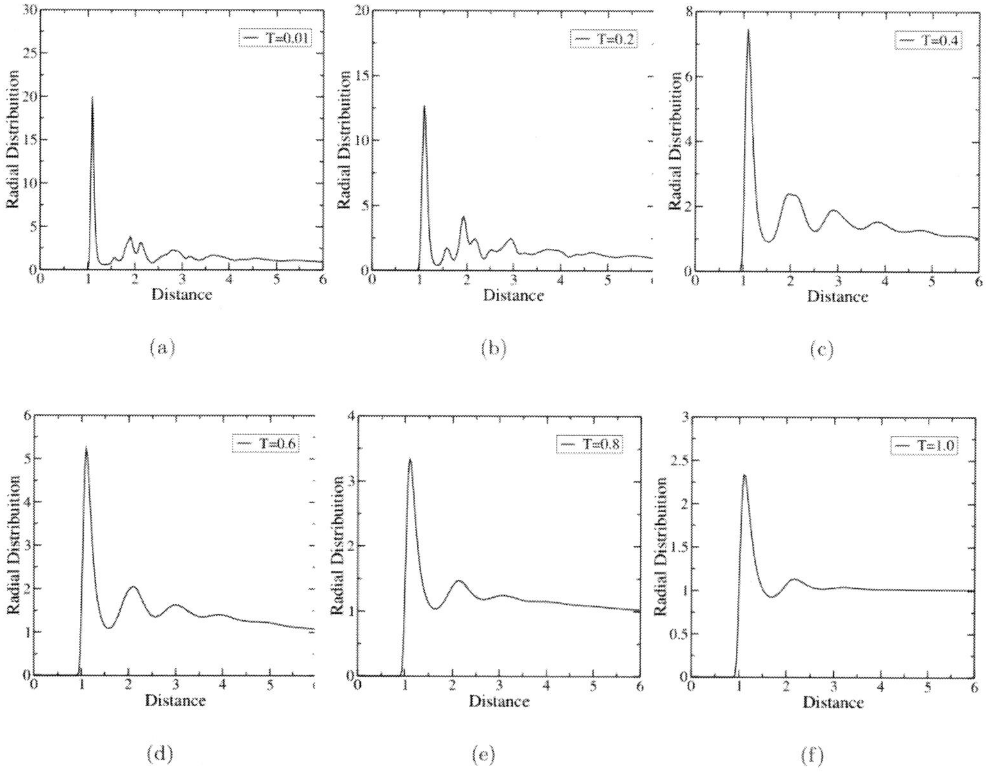

Figure 4. (Color online) The respective radial distribution functions for the configurations in Fig. 3.

3. Conclusions

The Shared Memory technique for the visualization of the simulations "on the fly" was presented. This technique for parallelization of the computation and the visualization of the simulations became accessible with the development of the multi-core technology as well as the expansion of the *RAM* memory. With the help of the *openGL* library the simulations on particle systems can be easily visualized. Consequently, the study of some physical phenomena like cluster formation, domain walls and vortex patterns in magnetic systems get another dimension. This integration between science and visualization is of paramount importance, either as an efficient debugger or as a tool to understand the simulated physical model.

Acknowledgments

The authors thank Dr. B. V. Costa for valuable discussions during the elaboration of this work.

References

[1] P. A. Fishwick *Web-based Simulation: Some Personal Observations*, Proceedings of the 1996 Winter Simulation Conference, vol. 96, pp.772 -779 1996

[2] C. Monserrata, U. Meierb, M. Alcaizb, F. Chinestac, M.C. Juana *A new approach for the real-time simulation of tissue deformations in surgery simulation*, Computer Methods and Programs in Biomedicine, Volume 64, Issue 2, February 2001, Pages 7785

[3] William Humphrey, Andrew Dalke, Klaus Schulten *VMD: Visual molecular dynamics*, Journal of Molecular Graphics, Volume 14, Issue 1, February 1996, Pages 3338

[4] M.W. Berry, *Massive Data Visualization*, Computing in Science &, Engineering, Vol. 1, No. 4, 1999, pp. 16-17

[5] Ananth Grama, George Karypis, Vipin Kumar, Anshul Gupta *Introduction to Parallel Computing*, 2nd ed. ,Pearson Education, Harlow, England.

[6] Peter Pacheco *An Introduction to Parallel Programming*, Elsevier, USA.

[7] Georg Hager, Gerhard Wellein *Introduction to High Performance Computing for Scientists and Engineers (Chapman & Hall/CRC Computational Science)*, CRC Press, Boca Raton, FL.

[8] Barbara Chapman, Gabriele Jost, Ruud van van der Pas *Using OpenMP: Portable Shared Memory Parallel Programming (Scientific and Engineering Computation)*, The MIT Press, Cambridge, Massachusetts.

[9] Eberth Correa, A. B. Lima and B. V. Costa, *Int. J. Mod. Phys.* C, Vol. 23, No. 4 (2012) 1250026.

[10] Dave Shreiner and Bill The Khronos OpenGL ARB Working Group, *OpenGL Programming Guide: The Official Guide to Learning OpenGL*, Addison-Wesley, Michigan, USA.

Journal of Physics: Conference Series **487** (2014) 012015 doi:10.1088/1742-6596/487/1/012015

Simulation and visualization of ion-implantation in diamond

Joan Adler, Amihai Silverman, Niv Ierushalmi, Anastassia Sorkin and Rafi Kalish

Technion - IIT, Haifa, Israel, 32000

E-mail: phr76ja@tx.technion.ac.il

Abstract.
We have explored aspects of ion implantation in diamonds with molecular dynamics and tightbinding atomistic simulations. Relevant experiments and their potential applications as well as our computer models and computational approaches are described. Our simulations have been designed to answer questions proposed by experimental researchers concerning optimal laboratory schedules for the preparation of samples with potential applications to diamond membranes and NV centers for quantum computers. Simulation and visualization of results enable us to peek inside samples where experimental techniques cannot tread. In order to provide the requisite Brazilian component a new connection between these models and bootstrap percolation is made.

1. Introduction

The Computational Physics group at the Technion carries out atomistic simulations in carbon systems, with an emphasis on geometric structure and vibrations, especially near defects and interfaces. Our simulations are made in close contact with several Technion experimental groups. Visualization is a major component of our studies to observe the system development, and we are often concerned with the structure deep inside the sample, which may not be accessible to non-destructive experiment.

After a brief introduction to the physics and chemistry of carbon allotropes, we will describe previous experiments and introduce some theoretical models that are relevant for their interpretation. Discussions of modelling with classical or quantum approximations or analyzing only initial and final states versus modelling the dynamics of the implantation process will be made. Early studies of damage in diamond will be reviewed. We will then recapitulate two recent calculations of processes involved in understanding experiments on samples prepared for applications relating to diamond membranes and NV (Nitrogen-Vacancy)centers. These are both designed to provide guidance to the experimental researchers about scheduling and temperature selections to achieve optimal sample quality. We will show how the use of animation aids in comphrehension. The situations described in this paper are connected with R. Kalish's experiments, but the computational group also collaborates with A. Hoffmann, M. Sheintuch and Y. Yaish on other carbon projects.

2. Background and previous studies

2.1. Carbon Allotropes

Carbon has several allotropes. One is diamond, a transparent insulator and another, graphite, an opaque conductor. The vastly different properties stem from their different atomic coordination and band structures. Transformation between the two is highly relevant for their applications - damaged industrial diamond loses its insulating properties and can become a conductor while damaged diamond windows, e.g. on spacecraft, can lose transparency. Understanding of the damage process leads to its avoidance or application, as appropriate.

2.2. Implantation experiments

Ion implantation is a tool to explore the stucture of materials, a means to create samples with specific structures in certain regions and a tool to introduce dopants. A classic example of the former is the Rutherford experiment that deduced the size of the nucleus by shooting α particles at gold foil and determining their angle of deflection. A classic example of the latter is the series of experiments in R. Kalish's laboratory [1] where diamond was transformed into a conductor by ion implantation.

2.3. Percolation theory

The percolation transition occurs when a two-component sample, one of whose components were previously disconnected has a changeover to a connected one. For example, in a mixture of randomly positioned glass and metal balls, when the concentration of metal balls increase from zero to 100% at some specific concentration of metal there will be a conducting path that spans the sample. In 1980 Kalish/Shapiro *et al.* [2] observed that regions of damaged diamond under implantation connect and insulating diamond becomes conducting carbon. A mean field percolation model was made and they conjectured the damaged regions are locally graphitic.

2.4. Static versus dynamic, classical versus quantum

Extensive calculations of energetics for samples with defects and interstitials are often made with the very best potentials, Density Functional Theory (DFT) approximations or even better wavefunction forms. Energy barriers are also commonly calculated. However full paths for sample transformations are very important in order to determine which experimental conditions, such as temperature or scheduling of annealing are optimal and these require true (molecular) dynamics which sometimes cannot be carried out with better than good classical or tightbinding potentials due to time/memory constraints.

2.5. Damage paths

Let us now consider the simplest case of irradiation. Atoms shot (Figure 1) into a diamond sample with sufficient energy make a damage cascade, and we asked what is the final step (Figure 2) of this cascade? In Figure 1 a cartoon of this situation is presented with the final state denoted by ?????? As shown in Figure 2 it was found [3] to be a split interstitial defect. This is a defect where two interstitial atoms share one lattice site. It was long known to be the most stable defect from calculations of the static energy, but the path to its creation was unclear previously. The two atoms at the centre of the defect are threefold coordinated (Figures 3 and 4) and the other atoms relax around it. An animation is shown in the supplementary material.

The next question is what happens when many atoms are fired towards the same region? As conjectured by Kalish/Shapiro, we found [4] that a locally graphitic region is created after annealing, see Figures 6 and 7. The image in Figure 4 was drawn with AViz, the other images from an older OpenGL code. In Figures 5, 6 and 7 we show only the 3-fold and 4-fold atoms, respectively (of the same sample) to enhance clarity.

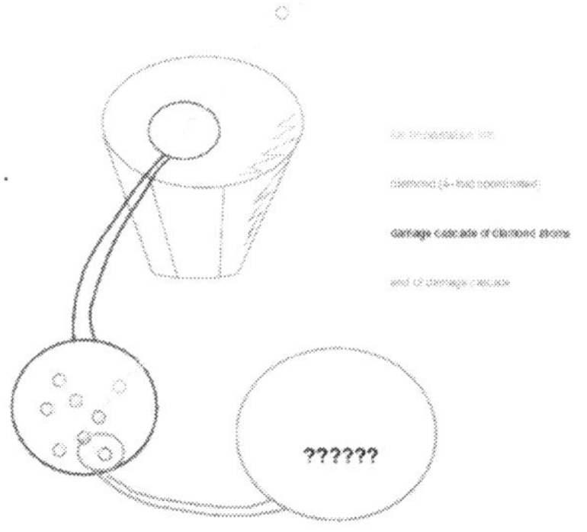

Figure 1. Before Saada, Adler and Kalish calculations.

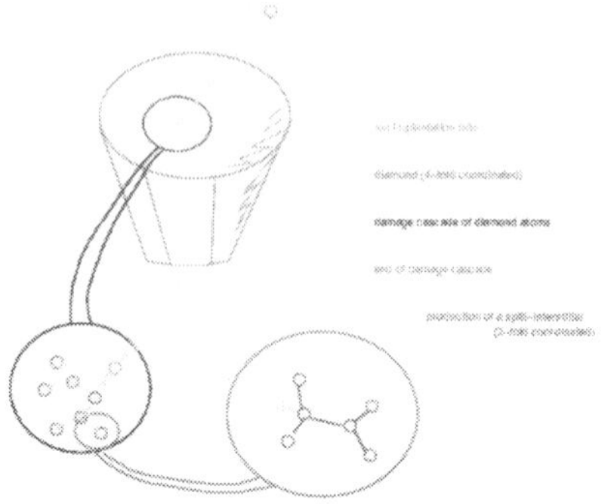

Figure 2. After our calculations.

Figure 3. Color key.

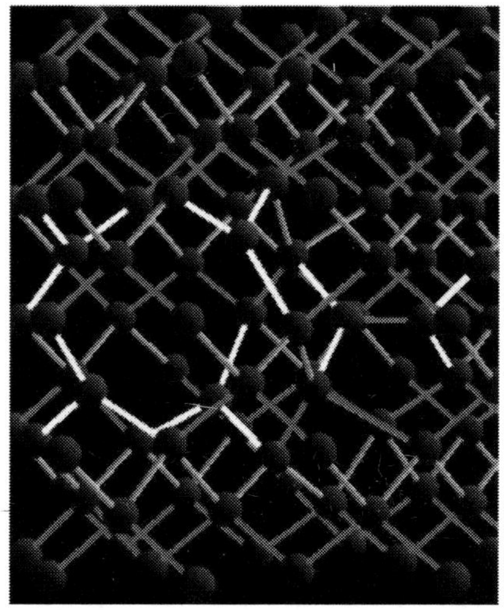

Figure 4. Formation of split interstitial.

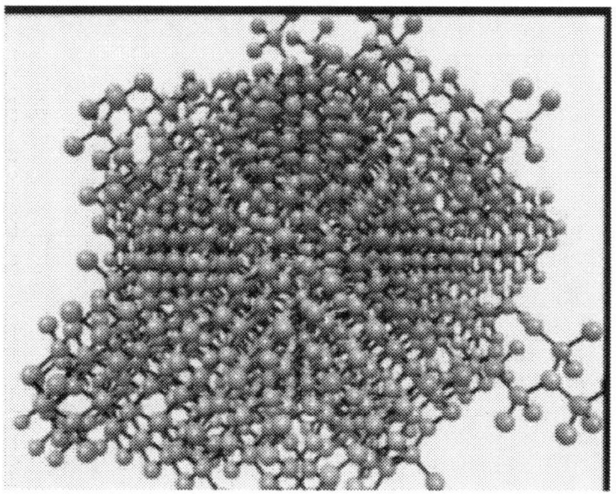

Figure 5. After irradiation with multiple atoms, and annealing, these atoms that have changed position. Red atoms 3-fold coordinated, blue atoms 4-fold.

We conclude this recapitulation with cartoons of before (Figure 8) and after (Figure 9) the Shapiro/Kalish calculations, envisioning multiple graphitic regions. Is this scenario realistic? Do the graphitic regions percolate? We expect that with sufficient irradiation, the regions should join and percolate. Explicit direct simulation of this has been on our "todo list" for nearly 30 years, but has been limited by computer resources so was always sidetracked by other projects. Hopefully it will be explored shortly now that we have improved parallel computing resources at the Technion.

2.6. Transformations

Irradiation causes local heating since there is an increase in kinetic energy. There are many ways to transform diamond to graphite and graphite to diamond, especially in the presence of hydrogen. For example - we can grow nanodiamonds (Figure 10) and nanographite (Figure 11). These images are from the thesis research of A. Sorkin and are described in [5]. The calculations were tightbinding MD with the OXON code. The transformations were often made by heating

Journal of Physics: Conference Series **487** (2014) 012015 doi:10.1088/1742-6596/487/1/012015

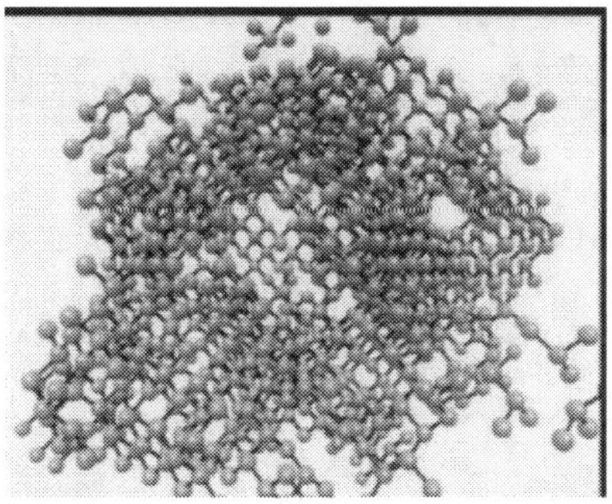

Figure 6. 4-fold coordinated atoms from Figure 5.

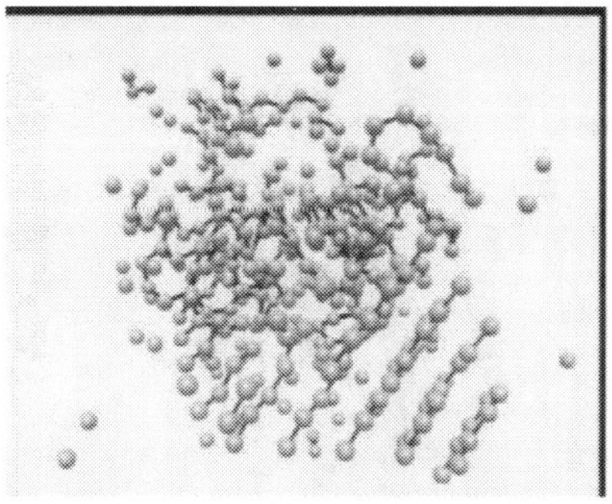

Figure 7. 3-fold coordinated atoms from Figure 5.

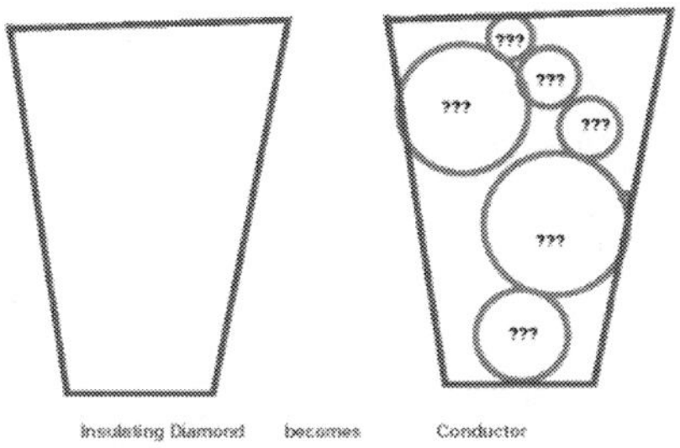

Figure 8. Before Shapiro/Kalish.

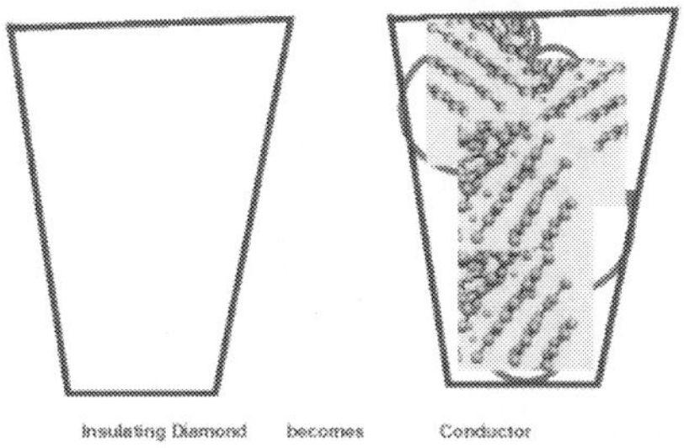

Figure 9. After Shapiro/Kalish.

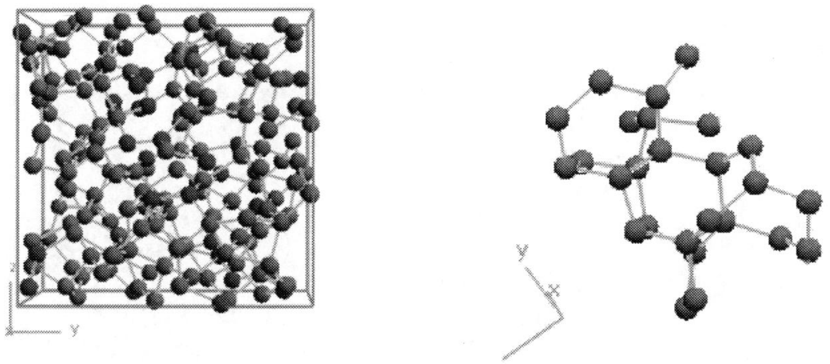

Figure 10. Initial sample before nanodiamond formation from [5] (left) and nanodiamond (right).

part of the samples while other parts were frozen, rather than by actual irradiation. The results depended on many factors and while we may claim that we were able to predict nanographite prior to its experimental identification, in fact we had seen this also much earlier in Saada et al[4]. Nanodiamonds such as ours were not identified earlier, possibly due either to insufficent equilibration, but more likely due to our improved methods for identifying the nanodiamond

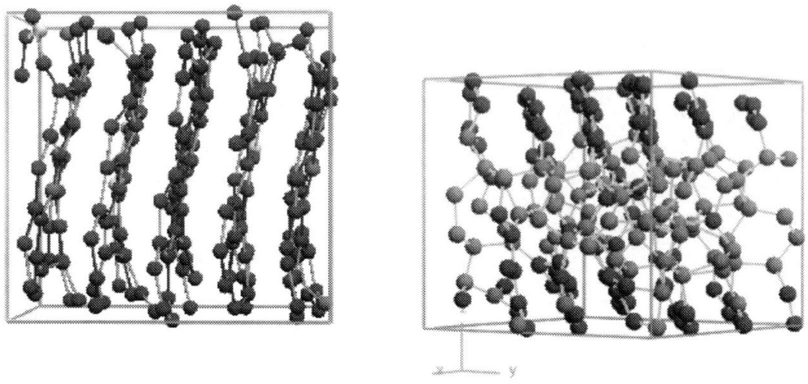

Figure 11. Nanographite formation from [5]. Intermediate cooling rate (right) and slow cooling rate (left). The pressure leads to flexed layers.

structures with our visualization tools. An animation of nanographite formation is shown in the supplementary material.

3. Diamond membrane creation

An overview of the preceeding section suggests that three relevant parameters for conversion between diamond and graphite are:

- Implantation energy
- Implantation temperature
- Post implantation annealing

Let us now consider the application of these to the fabrication of diamond membranes, wherein photonic crystals and other sub micro-sized optical devices can be realized. Many spintronic devices are based on specific optically active atomic structures in diamond, such as the nitrogen-vacancy center, and rely on the membranes performance, [6]. One promising approach for realizing such membranes is by creating a heavily damaged layer (rich in broken bonds) in diamond by ion implantation. Following annealing, this layer converts to graphite, which can be chemically removed, leaving a free-standing diamond membrane. Unfortunately, the optical properties of the exposed diamond surface (the diamond-vacuum interface) of such membranes currently are insufficient for high-quality photonic devices and so their performance is as yet, unsatisfactory. Thus, it appears that application of the current ion- implantation/graphitization methods do not yield the optical qualities that would enable the utilization of such lift-off methods to obtain high-quality waveguides and other nanosized optical devices. Understanding the properties of this interfacial layer and finding ways to minimize its detrimental effects on the reflectivity of photons impinging from within the membrane are of major importance. Some laboratory studies along these lines have recently been performed, but further guidance from simulation is desirable.

3.1. Surface smoothness

Since we have already explored transformation of diamond to graphite, it seemed natural to apply our simulation approach to deducing which combination of the three factors of implantation energy and temperature and of annealing temperature would provide the smoothest membrane. We deduced in a recent study [7] where different implantation and annealing conditions were simulated, that cold implantation, followed by high-temperature annealing leads to the creation of the sharpest diamond-etchable graphite interface, which should exhibit optimal optical properties among diamond membranes created by the implantation/graphitization method.

In this study we used a Brenner potential and Langevin thermostat and had 5120 atoms in each sample. We inserted 20 carbon atoms sequentially inserted at a random place between layers 11 and 12 and shot at an angle of 7 degrees with a Kinetic Energy of 350eV see Figure 12. We found damage that extended for about 20 (out of 40) layers. 10 sets of random locations were each done according to 3 schedules with a range of temperatures for schedules 2 and 3, as shown in Figure 13 We show the variation over samples with identical conditions, (in this case we plot the fraction of sp^2 bonds for each layer number, annealed at 2000 with schedule 2) except for the random seed in Figure 14. We observe that damage profile details vary for each set of initial random locations, but the overall picture is similar.

A sample with schedule 2, annealed at 2000 degrees is shown in Figure 16, where we color bonds that connect 4-fold coordinated sp^3 atoms in blue and 3-fold coordinated sp^2 ones in yellow. We also analyzed the slope of the sp^2 fraction in each layer and the results are shown in Figure 15. The sharpest is for the case shown in Figure 16. Another visual technique was to omit the 3-fold coordinated atoms completely. All these were based on the sample states at the end of the simulation, prior to the actual etching. A movie of therotation of the middle sample from Figure 17 is shown in the supplementary material.

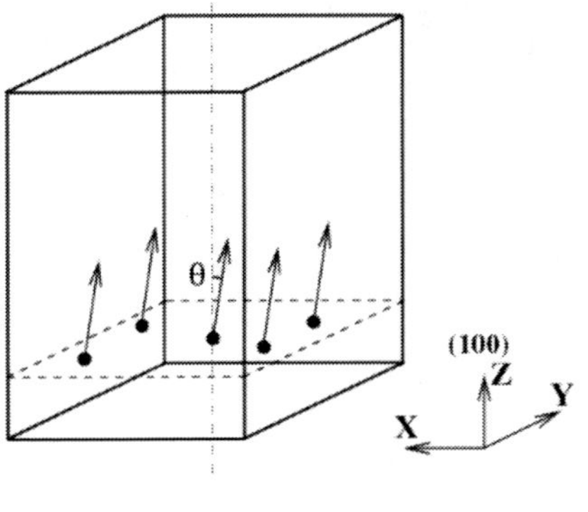

Figure 12. Model system.

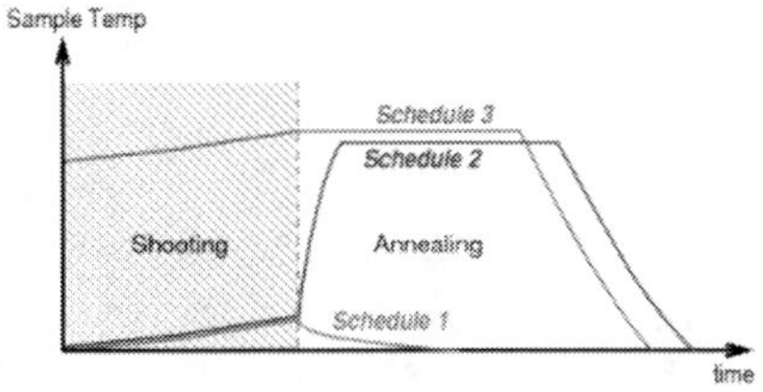

Figure 13. Shooting and annealing schedules.

3.2. Bootstrap percolation model for the etching process

In the etching process, atoms are sucessively removed as sp^3 diamond bonds are successively converted to clusters of weak etchable sp^2 graphitic bonds. This is reminiscent of a type of correlated percolation, known as bootstrap percolation (BP). In BP all sites with insufficient neighbours are iteratively removed [8], with say $m = 3$ BP having dangling bonds and isolated sites removed; i.e. only sites with 3 neighbours remain. BP leads to smoothing (lower fractal dimension) of samples and although proposed in a magnetic context there has been dearth of condensed matter applications, with possible exceptions being a series of studies of orientational order in molecular hydrogen[9]. Simulating the etching of the graphite from the membrane is a potential application! This application is a little different from the standard lattice based BP

Journal of Physics: Conference Series **487** (2014) 012015 doi:10.1088/1742-6596/487/1/012015

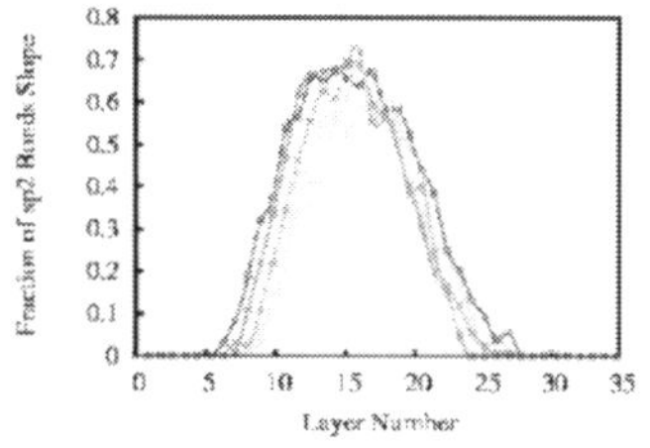

Figure 14. Sample variation.

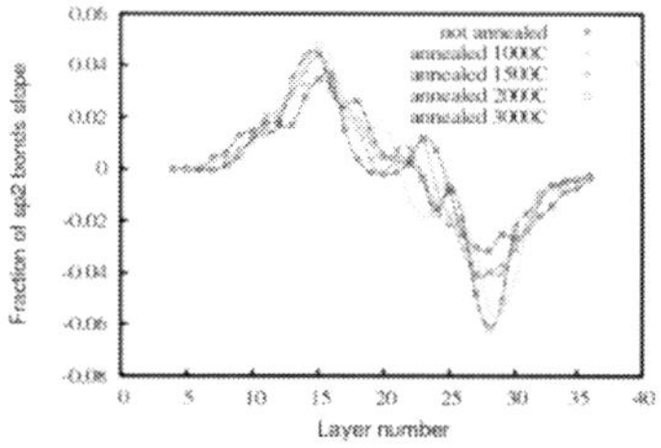

Figure 15. Slope of interface for different schedules.

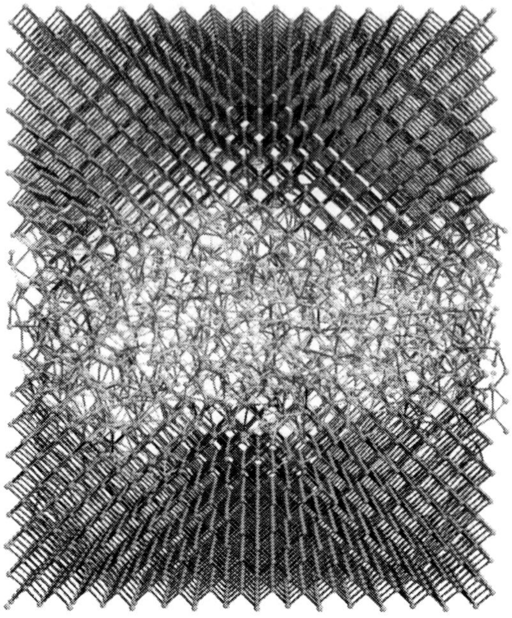

Figure 16. Sample after irradiation, showing 3-fold coordinated atoms in yellow.

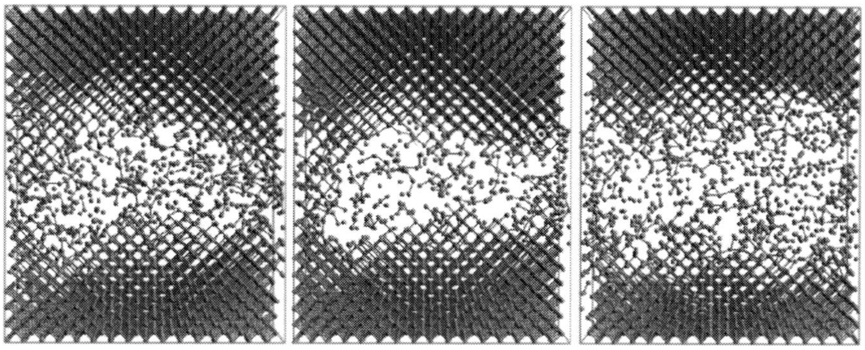

Figure 17. Three samples with schedules 1,2 and 3 (l. to r.) with bonds between 3-fold coordinated atoms removed.

toy models for two reasons. Firstly, the system is off-lattice and therefore a bond length has to be selected. Secondly, the bonding configuration of the site's neighbours has to be considered.

We compared the results of schedules 2 and 3 of [7] (called 1 and 2 on the website [10]). All atoms with less than 2 diamond neighbours within a distance of 1.55 angstrom were removed. The initial frame of one case is shown in Figure 18, the final cleaned upper surface in Figure 19 and the MATLAB construction to measure the fractal surface in Figure 20. A movie of the bootstrapping process is shown in the supplementary material. The MATLAB boxcounting routine gave results for the fractal dimension for schedule 2 of 2.2614 +/- 0.0034, and for schedule 3 of 2.2695 +/- 0.0038. This implies cold irradiation gives a smoother surface in agreement with [7]. More details are on the website [10].

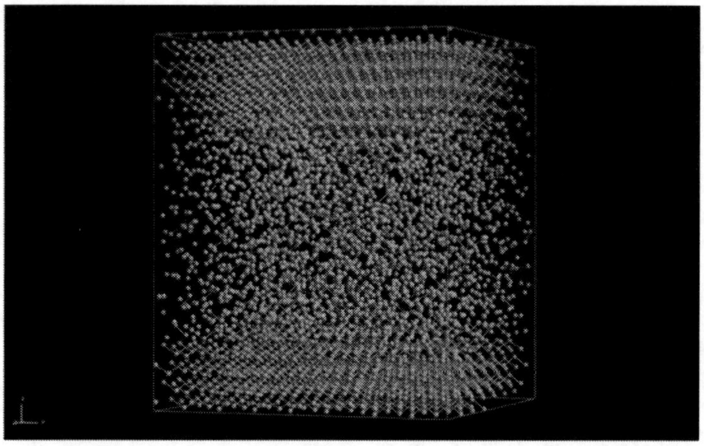

Figure 18. Sample after irradiation.

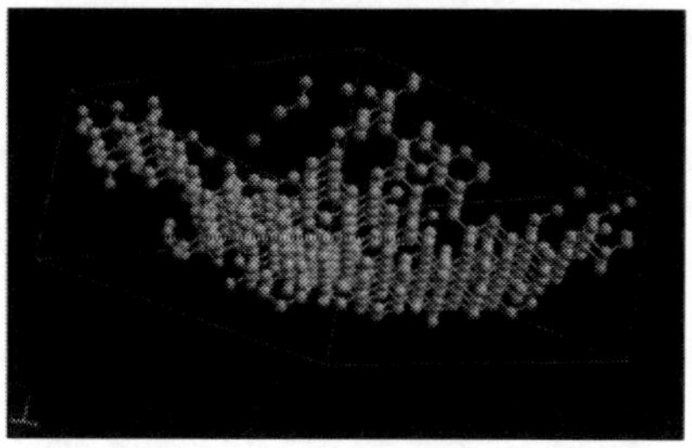

Figure 19. Final etched upper surface

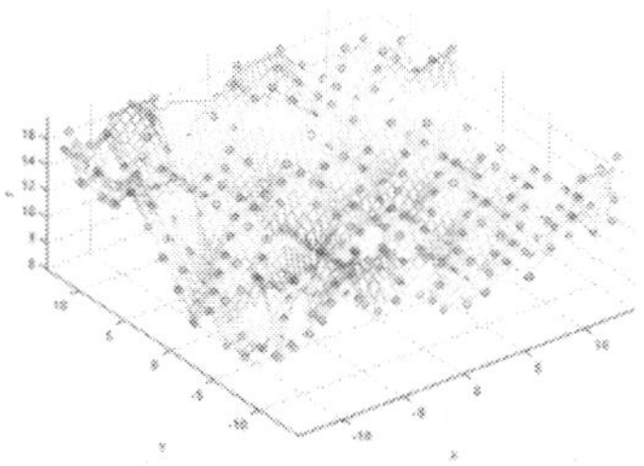

Figure 20. sample used for boxcounting to determine fractal dimension.

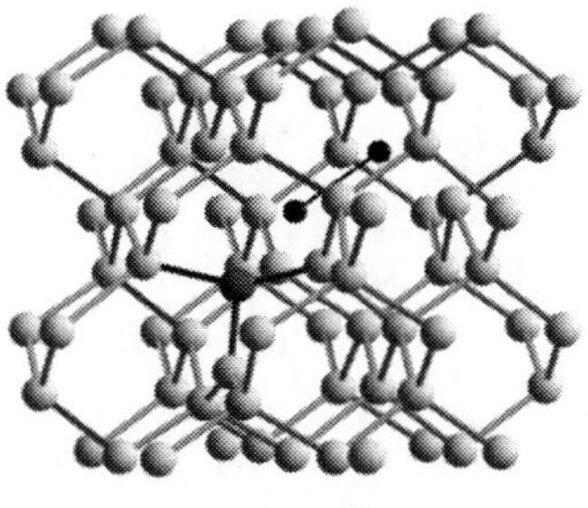

Figure 21. Nitrogen interstitial in blue (large ball) and two vacancies in black (small balls). Carbon atoms in yellow.

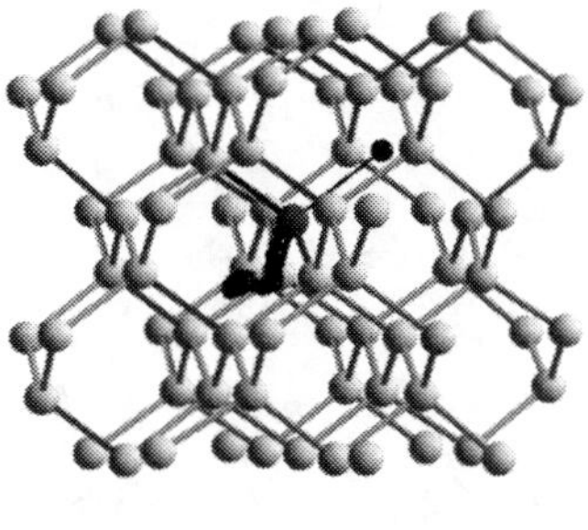

Figure 22. NV center after equilibration, successive locations of the Nitrogen atom are superimposed and can be viewed in a movie in the supplementary material.

4. NV center formation

Another process of interest for quantum computing applications is the creation of NV (Nitrogen-Vacancy) centers that are photoluminescent. In Rafi Kalish's laboratory research into optimal implantation techniques is underway. We have made a simulational study (both dynamics and energetics) with the dftb code to determine which schedules will optimize the realization of

NV (nitrogen-vacancy) centers in diamond. We found that when the nitrogen resides on an interstitial site, it fails to attract a vacancy, hence no NV center can be formed. However, if it occupies a split interstitial site and two vacancies reside in the second or third neighbor sites, an NV center will form following annealing at temperatures as low as 300°C and 650°C, respectively. This provides guidelines to experimentalists on how to increase the efficiency of NV formation in diamond. We were able to follow the dynamics of the creation process (as well as determine that other paths failed to create the center). In Figure 21 we show the initial sample and in Figure 22 the final state as well as the superimposed route. A movie is provided in the supplementary material, and on the website [11].

5. Conclusions

The special aspect of our approach is the close contact with the experimental laboratory. Experiments provide interesting problems to compute and simulations provide guidance to experimentalists. Our emphasis is on visualizing dynamic processes - simulation treads where microscopes cannot go.

Acknowledgments

Thanks to Alon Hoffman and the former students on carbon/diamond projects - David Saada and Yoel Koenka for collaboration over the years. Aernout van Enter is acknowledged for his updates on the progress of BP studies. The BP study of the membrane formation was carried out by Niv Ierushalmi as his project in the Computational Physics class; other projects from this class are described in [12].

References

[1] R. Kalish,"Nano-scale modification and doping of diamond:Interesting science and promising technology", Int. J. Nanotechnology, **6**, pp.691-703 (2009).

[2] R. Kalish, T. Bernstein, B. Shapiro and A. Talmi, "A percolation theory approach to the implantation induced diamond to amorphous-carbon transition", Radiation effects, **52**, pp.153-168 (1980).

[3] D. Saada, J. Adler and R. Kalish, "Transformation of the Diamond (sp^3) to Graphite (sp^2) bonds by ion-impact", International Journal of Modern Physics, C, **9**, 61-9 (1998).

[4] D. Saada, J. Adler and R. Kalish, "Computer simulation of damage in diamond due to ion-impact and its subsequent annealing", Phys. Rev. B, **59**, 6650-60 (1999).

[5] A. Sorkin, J. Adler and R. Kalish, "Computer simulations of damage due to passage of a fast heavy atom thru diamond", Phys. Rev B. **70**, 064110 (2004); A. Sorkin, J. Adler and R. Kalish, "Nucleation of diamond from liquid carbon under extreme pressures: Atomistic simulation", Phys. Rev. B **74**, 064115 (2006); A. Sorkin, R. Kalish and J. Adler, "Nanodiamond under pressure: Simulations ina the presence of Hydrogen", Phys. Rev. B., **78**, 155435 (2008).

[6] R. Kalish, "The role of ion implantation in the realization of spintronic devices in diamond", Nucl. Instr. Methods, B 272, 42 (2012); R. Kalish, "Ion iplaaantation in diamond; Doping and damaging", invited chapter in "Quantum information in diamond", edited by S. Prawer and I. Aharonovich, Woodhead Publishing, in press (2013).

[7] A. Silverman, J. Adler and R. Kalish "Diamond membrane surface after ion implantation induced graphitization for graphite removal:molecular dynamics simulation" Phys. Rev. B, **83** 155410 (2011).

[8] J. Adler and U. Lev, Brazilian Journal of Physics, **33** 641, (2003).

[9] J. Adler, R. G. Palmer and H. Meyer, "Transmission of order in some unusual dilute systems", Phys. Rev. Lett. **58**, 882-5 (1987).

[10] http://phelafel.technion.ac.il/∼ nivieru

[11] submitted, see also http://phjoan23.technion.ac.il/∼phr76ja/nv

[12] "How do I simulate problem X?", Joan Adler, Yaron Artzi, Liron ben Bashat, Tzipora Yael Izraeli, Meital Kreif, Ido Lavi, Alexander Leibenzon, Adam Lèvi, Itai Schlesinger, Elad Toledano, Uria Peretz, Yonatan Weisler and Alon Yagil, submitted to IOP Conference Series, (CCP2013).

The influence of local majority opinions on the dynamics of the Sznajd model

Nuno Crokidakis

Departamento de Física, PUC-Rio, Rio de Janeiro, Brazil

E-mail: nuno.crokidakis@fis.puc-rio.br

Abstract. In this work we study a Sznajd-like opinion dynamics on a square lattice of linear size L. For this purpose, we consider that each agent has a convincing power C, that is a time-dependent quantity. Each high convincing power group of four agents sharing the same opinion may convince its neighbors to follow the group opinion, which induces an increase of the group's convincing power. In addition, we have considered that a group with a local majority opinion (3 up/1 down spins or 1 up/3 down spins) can persuade the agents neighboring the group with probability p, since the group's convincing power is high enough. The two mechanisms (convincing powers and probability p) lead to an increase of the competition among the opinions, which avoids dictatorship (full consensus, all spins parallel) for a wide range of model's parameters, and favors the occurrence of democratic states (partial order, the majority of spins pointing in one direction). We have found that the relaxation times of the model follow log-normal distributions, and that the average relaxation time τ grows with system size as $\tau \sim L^{5/2}$, independent of p. We also discuss the occurrence of the usual phase transition of the Sznajd model.

1. Introduction

In the last thirty years the dynamics of agreement/disagreement has been extensively studied by physicists, which is nowadays an active branch of *Sociophysics* [1, 2, 3]. Agreement is one of the most important features of social dynamics. Everyday life presents many situations in which it is necessary for a group to reach shared decisions. Agreement makes a position stronger, and amplifies its impact on society. Another interest in social phenomena is its theoretical motivation. In fact, several models present long-range correlations, power-law behavior and order-disorder phase transitions, among other typical features of physical systems [1, 2, 3].

Among the most studied opinion models in the last years we highlight the Galam models [4], the voter model [5] and the Sznajd model [6]. In special we are interested in the Sznajd model, that was initially defined on an one-dimensional ring, but was also considered on regular lattices in two and three dimensions, on complex networks and other topologies, and was applied to several practical problems like finance, marketing and politics (for a review, see [7]).

Despite the above applications of the Sznajd model, it does not consider important social features. The dynamics of social relationships in the real world shows a large number of details that are commonly neglected in some theoretical models. In order to introduce more realistic features, we have considered in this work a convincing-power mechanism that limits the capacity of persuasion of the agents in the Sznajd model, as discussed in [8]. In addition, we have considered that not only groups of four agents sharing the same opinion can persuade their neighbors, but also groups with a local majority opinion can be persuasive in the opinion formation process. It is expected that the inclusion of such ingredients in the Sznajd model turns it closer to a real social system, where not only the number of individuals with same opinion matters, but also their persuasiveness. Furthermore, it has been discussed that a local majority opinion can be considerably influent on the dynamics of public debates [4].

This work is organized as follows. In Section 2 we present the model and define their microscopic rules. The numerical results as well as the finite-size scaling analysis are discussed in Section 3. Finally, our conclusions are presented in Section 4.

2. Model

The model considered in this paper is an extension of a recente work [8]. We have considered the Sznajd model defined on a square lattice with linear size L and periodic boundary conditions [9]. Each agent on a given lattice site i carries an Ising-like variable $s_i = \pm 1$, representing the two possible opinions (yes or no, for example) in a given subject. The initial configuration of opinions is given by a fraction or density d $(1-d)$ of $s = +1$ $(s = -1)$ opinions. Furthermore, an integer number (C) labels each player and represents its *convincing power* across the community. The convincing power is introduced as a score which is time dependent. The agents start with a random distribution of the C values, and during the time evolution the convincing power of each agent can change according to its capacity of persuasion, following the model's rules. One time step in our model is defined by the following rules:

(i) We randomly choose a 2×2 plaquette or group of four neighbors.

(ii) If all four center spins are parallel, we calculate the average convincing power $\bar{C}$ of the plaquette:

$$\bar{C} = \frac{1}{4}\sum_{i=1}^{4} C_i \ ,$$

where each term C_i represents the convincing power of one of the plaquettes' agents.

(iii) Then, we compare the average convincing power $\bar{C}$ with the convincing power of each one of the 8 sites neighboring the plaquette. Thus, given a plaquette neighbor j with convincing power C_j, this individual j will follow the plaquette opinion if $\bar{C} > C_j$. In this case, the convincing power of each agent in the 2×2 group is increased by 1.

(iv) If not all four spins are parallel, we verify if there is a majority opinion in the group (3 up/1 down spins or 1 up/3 down spins). In this case, we apply the above rule with probability p, i.e., we calculate the average convincing power $\bar{C}_{\mathrm{maj}}$ of the agents with the majority opinion, and compare $\bar{C}_{\mathrm{maj}}$ with the convincing power of each one of the 8 sites neighboring the plaquette. For each persuasion, the convincing power of each agent with majority opinion in the 2×2 group is increased by 1.

Thus, the convincing powers of the agents neighboring a given group act in the dynamics as a conviction, in a way that if a given plaquette neighbor j has a convincing power $C_j > \bar{C}$ he/she will not be persuaded to change opinion. In this case, the above-mentioned rules define a competition among a group of agents sharing the same opinion and the conviction of the agents neighboring the group. In addition, we will consider also the influence of the majority opinion in the group. In this case, with probability p an opinion which is the majority one in a given group may be adopted by the group's neighbors, since this local majority opinion is shared by a high convincing-power group.

The objective of the agents in this dynamics is to convince their neighbors of their opinion. One can expect that, if a certain group of agents convinces many other agents, the persuasion

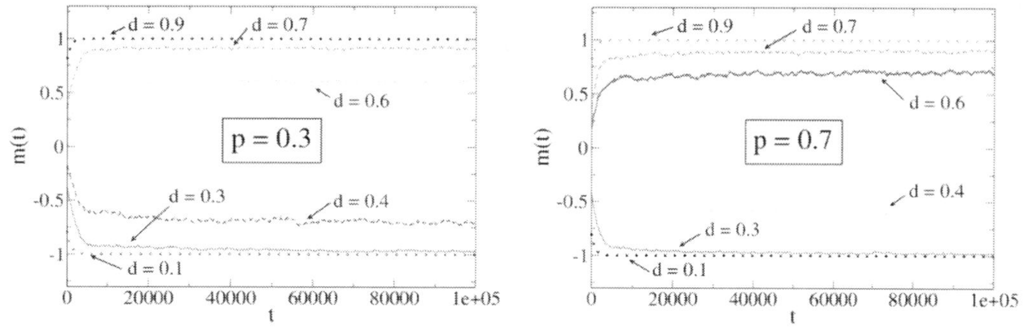

Figure 1. Time evolution of the magnetization per spin for $L = 53$, some values of the initial density d of $s = +1$ opinions and two different values of the probability p, namely $p = 0.3$ (left side) and $p = 0.7$ (right side). We can observe steady states with $|m| = 1$ (consensus) as well as with $|m| < 1$ (no consensus). Each curve is a single realization of the dynamics.

ability of each agent in the group increases. Thus, the inclusion of the convicing powers in the model may capture this real-world characteristic. In addition, as considered in Galam's models [4], a local majority opinion in a discussion group considerably affects the dynamics of a public debate. For the limiting case $p = 0.0$, we recover the model studied in [8].

3. Numerical Results

In the simulations, the initial values of the agents' convincing powers follow a gaussian distribution centered at 0 with standard deviation $\sigma = 5$ [1]. We can start studying the time evolution of the quantity analogous to the "magnetization per spin", i.e.,

$$m(t) = \frac{1}{N}\sum_{i=1}^{N} s_i \ , \tag{1}$$

where $s_i = \pm 1$ and $N = L^2$ is the total number of agents on the square lattice. We exhibit in Figure 1 the magnetization per spin m as a function of the number t of time steps for $p = 0.3$ (left side) and $p = 0.7$ (right side) and some values of the initial density d of $s = +1$ opinions. Observe that the time needed for the system to reach the steady states does not vary considerably from one case to another. One can also see in Figure 1 that situations with $|m| < 1$ in the long-time limit (steady state) emerge spontaneously from the dynamics, as in the $p = 0.0$ case [8]. This situation does not occurs in the original two-dimensional Sznajd model, where only absorbing states with $|m| = 1$ (consensus states) are obtained in the steady states [9]. In our case, such full-consensus states with all spins pointing up or down are possible only for large and small values of d, respectively, as we can also see in Figure 1. Thus, the consideration of the influence of a local majority opinion does not favor consensus states, usually related to dictatorships [9], making stable steady states with $|m| < 1$, that can be related to a democracy. This makes the Sznajd model more realistic. One can also see in Figure 1 that there are transitions between a phase where the full-consensus states with $|m| = 1$ are always reached and another phase where this consensus never occurs. This fact is more clear in Figure 2, where we exhibit the time evolution $m(t)$ for a large initial density of +1 opinions ($d = 0.9$) and for two distinct values

[1] As discussed in [8], the variation of σ does not affects quantitatively the results of the model, which also occurs in the present formulation.

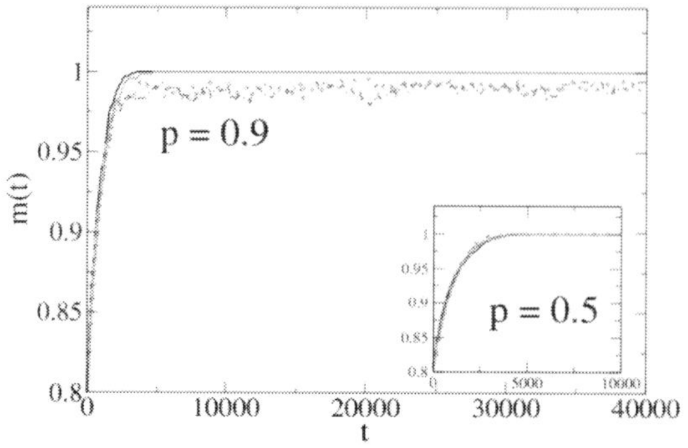

Figure 2. Time evolution of the magnetization per spin for $L = 53$, $d = 0.9$ and two distinct values of p, namely $p = 0.9$ (main plot) and $p = 0.5$ (inset). We have considered 5 independent simulations for each value of p. One can see that even for a large initial density of +1 opinions, the increase of the influence of groups with a majority opinion does not necessarily lead the system to the absorbing state with $m = 1$.

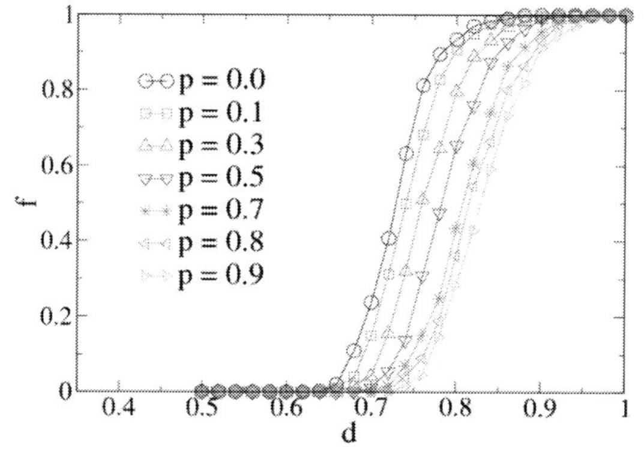

Figure 3. Fraction f of samples which show all opinions +1 when the initial density d of +1 opinions is varied in the range $0.5 \leq d \leq 1.0$, for $L = 73$ and typical values of the probability p. One can observe the usual phase transition from $f = 0$ to $f = 1$ at values $d_c(p)$, at least for $p < 0.8$.

of p, namely $p = 0.5$ and $p = 0.9$. In this case, the increase of the influence of groups with a majority opinion (i.e., the increase of the probability p) leads the system to steady states with $|m| < 1$. These results suggest that the usual phase transition of the Sznajd model may not occurs in our model for large values of p. We will analyze this phase transition in more detail in the following.

In order to better study the phase transition, we simulated the system for different probabilities p, and we measured the fraction f of samples which show all opinions +1 when the initial density d of +1 opinions is varied in the range $0.5 \leq d \leq 1.0$. In other words, this quantity f gives us the probability that the population reaches consensus, for a given value of the pair (d, p). The numerical results for f as a function of d for $L = 73$ and typical values

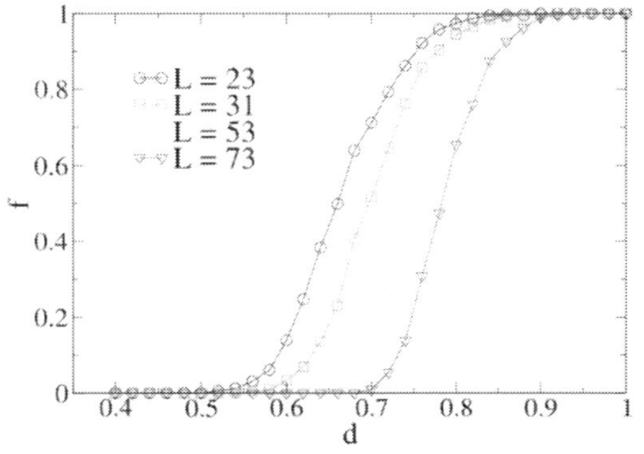

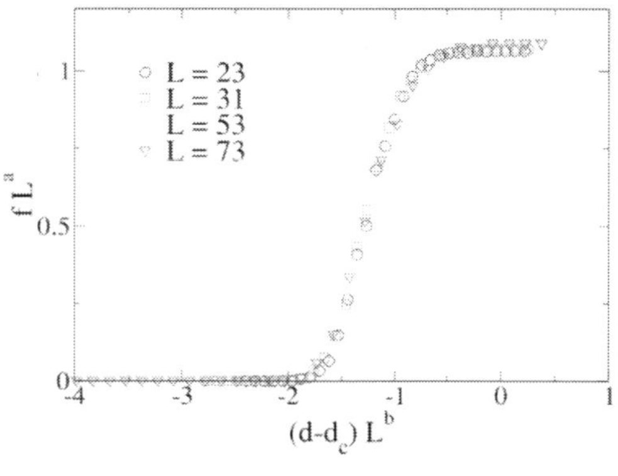

Figure 4. Fraction f of samples which show all opinions $+1$ when the initial density d of $+1$ opinions is varied in the range $0.4 \leq d \leq 1.0$, for $p = 0.5$ and some values of the lattice size L (left side). It is also exhibited the corresponding scaling plot of f (right side). The best collapse of data was obtained for $a = 0.02$, $b = 0.47$ and $d_c = 0.95$.

of p are shown in Figure 3. Observe that for values of p near 0.8 the system reaches the value $f = 1.0$ only for $d \to 1.0$, which suggests that there is no phase transition for large values of the probability p. For the formal characterization of the transition, we have to simulate the system for different lattice sizes L. We have considered 1000 samples for $L = 23$ and $L = 31$, 700 samples for 53 and 500 samples for $L = 73$. An example is exhibited in Figure 4 (left side), for $p = 0.5$. In order to locate the critical point, we performed a finite-size scaling (FSS) analysis, based on the standard FSS equations [8],

$$f(d, L) = L^{-a}\, \tilde{f}((d - d_c)\, L^{-b}) \;, \tag{2}$$

$$d_c(L) = d_c + a\, L^{-b} \;. \tag{3}$$

The result is shown in Figure 4 (right side), and we have found that

$$d_c(p = 0.5) = 0.95 \pm 0.01 \;, \tag{4}$$

in the limit of large L, with exponents $a \approx 0.02$, $b \approx 0.47$. Notice that the values of the critical exponents are similar to the ones observed in the $p = 0.0$ case [8]. Nonetheless, the value of d_c is greater than the critical density for $p = 0.0$, for which the numerical result is $d_c(p = 0.0) = 0.88 \pm 0.01$ [8]. This result indicates that an important effect of the influence of local majority opinions is to shift the critical point for higher values, but the universality class of the model remains the same for different values of the probability p.

In order to estimate the critical points $d_c(p)$ for other values of p, we repeated the above-discussed FSS analysis. The results suggest that we have $d_c(p) < 1.0$ (within error bars) for $p \leq 0.7$, which indicates the occurrence of a phase transition, with similar exponents $a \approx 0.02$, $b \approx 0.47$, which confirms the universality class for distinct values of p. On the other hand, the best collapse of data for $p > 0.7$ was obtained for $d_c = 1.0$. In this case, we can not affirm if the phase transition occurs. Taking into account the critical densities $d_c(p)$ calculated for typical values of p, we exhibit in Figure 5 the phase diagram of the model in the plane d versus p. The points are the critical densities $d_c(p)$ estimated from the FSS analysis of the numerical results, whereas the dashed line is just a guide to the eyes. Considering the thermodynamic limit

Journal of Physics: Conference Series **487** (2014) 012016 doi:10.1088/1742-6596/487/1/012016

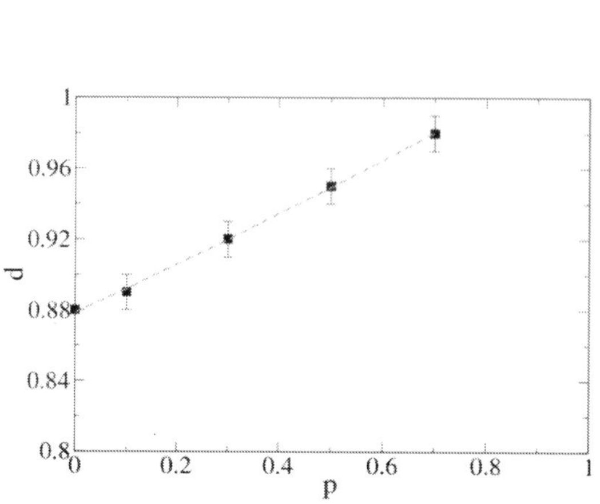

Figure 5. Phase diagram of the model in the plane d versus p. The points are numerical estimates of the critical densities $d_c(p)$, whereas the dashed line is just a guide to the eyes. In the region above the curve the system always reaches consensus with all opinions +1, whereas in the region below the curve this kind of consensus never occurs.

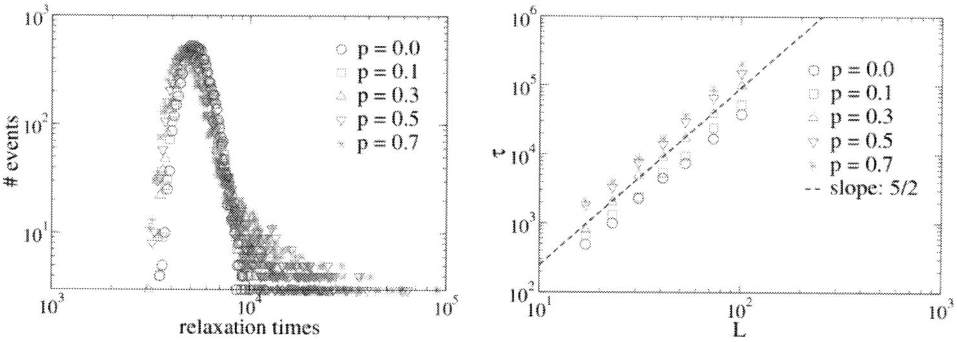

Figure 6. Histogram of the relaxation times for $L = 53$ and $d = 0.9$ (left side). It is also exhibited the log-log plot of the average relaxation time τ as a function of the lattice size L (right side). The dashed line has slope 5/2.

$L \to \infty$, the critical points d_c separate a region where the full-consensus states with all opinions +1 (or all spins up) always occur, for each realization of the dynamics (above the frontier), from a region where this kind of consensus never occurs (below the curve).

We have also studied the relaxation times of the model, i.e., the time needed to find all the agents at the end having the +1 opinion. The distribution of the number of sweeps through the lattice, averaged over 10^4 samples, needed to reach the fixed point is shown in Figure 6 (left side), for typical values of the probability p. We can see that the relaxation time behavior is compatible with a log-normal distribution for all values of the probability p, which corresponds to the parabolas in the log-log plot of Figure 6 (left side). In addition, the numerical results suggest that the relaxation times are not substantially affected by the probability p in the central region (near the peaks), which is in agreement with the above qualitative discussion based on the results shown in Figure 1. The difference appears in the tail of the distribution, that presents larger relaxations times for increasing values of p. These tails are responsable for the difference among the average relaxations times τ for a given value of the lattice size L, as one can see in Figure 6 (right side). The data are plotted in a log-log scale, and we can verify a power-law relation between those quantities in the form

$$\tau \sim L^{5/2} \; . \tag{5}$$

Notice that the above exponent is the same for all values of p, including the special case $p = 0.0$ [8], and thus it is not affected by the influence of local majority opinions.

4. Final Remarks

In this work, we have analyzed the consequences of the introduction of two distinct mechanisms in the Sznajd model. The first one is the convincing power, that is introduced as a score (C) which is time dependent. The agents start with a gaussian distribution of the C values, and during the time evolution, the convincing power of each agent changes according to its capacity of persuasion, following the model's rules. The second mechanism is the influence of local majority opinions. In this last case, if there is a local majority opinion in a group of 4 agents (3 up/1 down spins or 1 up/3 down spins), with probability p these agents sharing the majority opinion can persuade their neighbors to adopt the mentioned opinion. We analyzed the influence of these two ingredients on the dynamics of the Sznajd model and on the opinion spreading.

A first consequence of the presence of those ingredients is the absence of consensus (all opinions +1 or all spins up) for a wide range of parameters. These states are obtained only for very small or very large initial densities d of +1 opinions. In this sense, this formulation of the Sznajd model appears to be more realistic, since consensus states are usually related to dictatorships. On the other hand, states with magnetization per spin $|m| < 1$ are desirable since they represent democracy-like situations [8, 9]. A second effect of the influence of the local majorities is that the critical density of up spins $d_c(p)$ shifts for higher values with increasing values of p. In addition, the usual phase transition of the Sznajd model occurs for all values of $p \leq 0.7$, but for $p > 0.7$ we can not affirm if that phase transition occurs. In the cases where the transition occurs, the critical exponents are universal and they are independent of p.

As in the standard 2D Sznajd model [9], we found a log-normal distribution of the relaxation times to consensus. In addition, the average relaxation time τ depends on the lattice size in a power-law form, $\tau \sim L^{5/2}$, which is also independent of p.

Acknowledgments

The author acknowledges financial support from the Brazilian funding agencies FAPERJ and CAPES.

References

[1] Galam S 2012 *Sociophysics: A Physicist's Modeling of Psycho-political Phenomena* Springer (Berlin).
[2] Castellano C, Fortunato S and Loreto V 2009 *Rev. Mod. Phys.* **91** 591.
[3] Stauffer D 2013 *J. Stat. Phys.* **151** 9.
[4] Galam S 2008 *Int. J. Mod. Phys.* B **19** 409.
[5] Krapivsky P, Redner S and Ben-Naim E 2010, *A kinetic view of statistical physics* Cambridge University Press (Cambridge).
[6] Sznajd-Weron K and Sznajd J 2000 *Int. J. Mod. Phys.* B **11** 1157.
[7] Sznajd-Weron K 2005 *Acta Phys. Pol.* B **36** 2537.
[8] Crokidakis N, Forgerini F L 2010 *Phys. Lett.* A **374** 3380.
[9] Stauffer D, Sousa A O and de Oliveira S M 2000 *Int. J. Mod. Phys.* C **11** 1239.

Journal of Physics: Conference Series **487** (2014) 012017 doi:10.1088/1742-6596/487/1/012017

Normal-state correlated electronic structure of tetragonal FeSe superconductor

L. Craco[1], **M.S. Laad**[2], **S. Leoni**[3]

[1] Instituto de Física, Universidade Federal de Mato Grosso, 78060-900, Cuiabá, MT, Brazil
[2] Max Planck Institute for the Physics of Complex Systems, 01187 Dresden, Germany
[3] School of Chemistry, Cardiff University, Cardiff, CF10 3AT, UK

E-mail: `lcraco@fisica.ufmt.br`

Abstract. Tetragonal FeSe, a prototype iron-chalcogenide superconductor, shows signatures of a strange incoherent normal state. Motivated thereby, we use LDA+DMFT to show how multi-band correlations generate a low-energy pseudogap in the normal state, giving an incoherent metal in good semi-quantitative agreement with observations. Anomalous responses in the normal state, including orbital-dependent effective mass enhancement and photoemission lineshape, are consistently understood.

1. Introduction

High temperature superconductivity in the Iron pnictides and chalcogenides compounds [1] is the latest surprise among a host of correlated electron materials. While unconventional superconductivity sets in close to the border of a frustration-induced [2] striped-spin-density-wave state with doping in the so-called 1111-pnictides, no magnetic long range order is seen in the tetragonal phase of Iron Selenide (FeSe) [3] and $FeSe_{1-x}Te_x$ [4], labelled 11 systems, for small x in ambient conditions. Undoped FeSe exhibits superconductivity with T_c =9 K: upon tuning the carrier concentration of single-layer films T_c rises to 65 K [5]. Superconductivity is sensitive to stoichiometry - minute non-stoichiometry in $Fe_{1+y}Se$ destroys the superconducting state [6]. Unconventional superconductivity at $T_c = 34$ K is even observed in the high pressure *orthorhombic* structure in FeSe [7] in contrast to the 1111-pnictides, where it is stable in the tetragonal structure. Interestingly, a two-step increase in T_c as a function of pressure (with a large dT_c/dP beyond $P_{c1} = 1.5$ GPa) is observed [8]. In contrast, superconductivity in FeSe is suppressed under tensile strain [9]. Moreover, extant experiments for the normal state show electron correlation fingerprints. Photoemission (PES) experiments [10, 11] show evidence of an incoherent, pseudogapped metallic state [10] in FeSe, instead of a narrow Landau quasiparticle peak at the Fermi level, E_F. *Ab initio* band structure calculations [12] compare poorly with PES data, as is checked by direct comparison (see below). In addition, the ultrahigh-resolution PES spectra show a low energy kink at ≈ 8 meV [11]. Finally, an ARPES [13] study shows appreciable, orbital-dependent effective mass enhancement $(16 - 21)$ in the normal state of $FeSe_{0.42}Te_{0.58}$, directly testifying sizable correlations in this system. As in 1111-compounds [14], the kink in PES sharpens with cooling, and evolves smoothly across T_c. Depending upon x, superconductivity in $Fe(Se_{1-x}Te_x)$ either arises from an insulator-like normal state, or from a bad metal with $\rho_{dc}(T) \propto T$ [15]. Finally, a minute amount of alloying by Cu drives FeSe

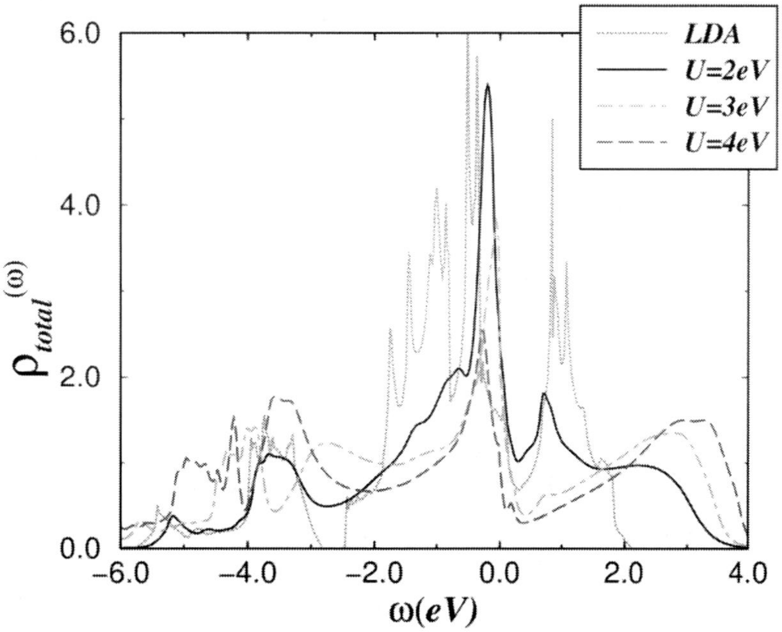

Figure 1. Comparison between the LDA (dotted) and LDA+DMFT (solid, dot-dashed and long-dashed) density-of-states (DOS) for the Fe d-orbitals in FeSe. Large-scale transer of spectral weight from low energy to the Hubbard bands with increasing U is visible. Also clear is the destruction of the low-energy Fermi liquid (FL) quasiparticle peak at $U = 4$ eV.

to a Mott-Anderson insulator [16]. Thus, FeSe is close to a metal-insulator transition, i.e, to Mottness [17]. Needless to say, a proper microscopic understanding of the coupled orbital-spin [18] correlations manifesting in such anomalous behavior in Fe(Se,Te) systems is a basic prerequisite for understanding how superconductivity emerges from such a normal state.

In this work we undertake a systematic local-density approximation plus dynamical mean-field theory (LDA+DMFT) [19] study of tetragonal FeSe. Sizable electronic correlations are shown to be necessary for gaining proper insight into the anomalous normal state responses in this system. Good semi-quantitative agreement with PES [10] supports our description.

2. Results and discussion

In our numerical simulation we start with the tetragonal (space group: $P4/nmm$) structure of FeSe with lattice parameters derived by Hsu *et al.* [20]. One-electron band structure calculations based on local-density-approximation (LDA) were performed for FeSe using the linear muffin-tin orbitals (LMTO) [21] scheme. Our LDA results for the total density of states (DOS) is shown in Fig. 1 (dotted line). Similar total DOS were also obtained by other groups [12], showing that the electronic states relevant to Fe-superconductors are Fe d-band states. As found in previous calculations, the Fe-d bands hybridize with Se-p bands around -3.8 eV, giving rise to a small, separated band of d character below 3 eV binding energy. Interestingly, the resulting "gap" at high energy is not seen in PES experiments [10, 11], which show only a broad continuum in this energy range. As discussed below, this discrepancy is resolved by dynamical spectral weight transfer (SWT) which originates from sizable electronic correlations in FeSe.

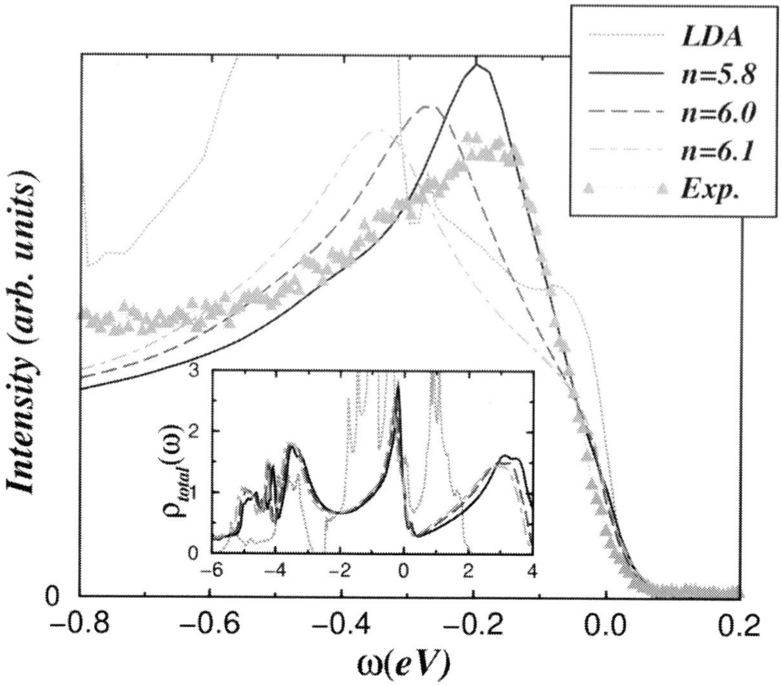

Figure 2. Comparison between the LDA+DMFT result for FeSe and angle-integrated photoemission (PES, triangles) [10]. Good semiquantitative agreement is seen for $n = 5.8$. In particular, the low-energy energy spectrum (up to 0.1 eV binding energy) and the peak at -0.17 eV in PES is resolved in the DMFT spectrum with $U = 4.0$ eV and $J_H = 0.7$ eV. (The inset shows the total LDA+DMFT spectral functions. LDA result is shown for comparison.)

Though LDA provides reliable structural information on a one-electron level, it generically fails to capture the ubiquitous dynamical correlations in d-band compounds, and so cannot access normal state incoherence in d-band systems. Combining LDA with dynamical-mean-field-theory (DMFT) is the state-of-the-art prescription for remedying this deficiency [19]. Within LDA, the one-electron part for tetragonal FeSe is $H_0 = \sum_{\mathbf{k},a,\sigma} \epsilon_a(\mathbf{k}) c^{\dagger}_{\mathbf{k},a,\sigma} c_{\mathbf{k},a,\sigma}$, where $a = x^2 - y^2, 3z^2 - r^2, xz, yz, xy$ label the diagonalized, five d bands. In light of the correlation signatures cited above full, multi-orbital (MO) Coulomb interactions must be included. These constitute the interaction term, which reads $H_{int} = U \sum_{i,a} n_{ia\uparrow} n_{ia\downarrow} + U' \sum_{i,a\neq b} n_{ia} n_{ib} - J_H \sum_{i,a\neq b} \mathbf{S}_{ia} . \mathbf{S}_{ib}$. To pinpoint the relevance of MO electronic interactions in the system, we present LDA+DMFT results for $U = 2, 3, 4$ eV, $U' = U - 2J_H$, and fixed $J_H = 0.7$ eV. In this sense, our study is not *ab initio*, but should be looked upon as a realistic correlated model and numerical simulation for FeSe. To solve the MO-DMFT equations, we use the MO iterated-perturbation-theory as an impurity solver [22].

Fig. 1 shows how LDA+DMFT modifies the LDA band structure. MO dynamical correlations arising from U, U' and J_H lead to spectral weight redistribution over large energy scales and the formation of lower- (LHB) and upper-Hubbard (UHB) bands. As seen, the UHB at 2.4 eV for $U = 2$ eV (and, $U' = 0.6$ eV) moves to higher energies with increasing U. The LHB is not clearly resolved for $U \leq 2$ eV. Indeed, we observe a relatively sharp and quasi-coherent

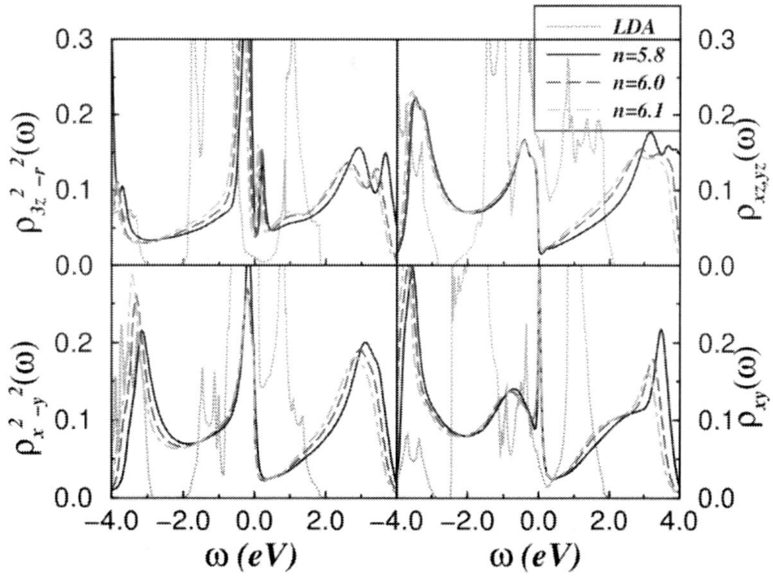

Figure 3. Orbital-resolved LDA (dotted) and LDA+DMFT (with $U = 4.0$ eV, $U' = 2.6$ eV and $J_H = 0.7$ eV) DOS for the Fe d-orbitals in FeSe for three doping values. Large-scale dynamical spectral weight transfer occouring hand-in-hand with orbital selective incoherence is visible.

low-energy peak, with a prominent shoulder feature instead of the LHB at $\omega \simeq -1.0$ eV. Similar features are visible in other results [23] for similar U values. Correlation effects, however, become more visible at $U \geq 3$ eV. In contrast to the $U = 2$ eV result, a LHB at 2.8 eV binding energy is clearly resolved with $U = 3$ eV. With increasing U, the LHB is shifted toward energies where the Se-p bands occur in the LDA: this superposition of the pd-band and LHB for $U = 4$ eV makes difficult to observe the LHB experimentally. Fig. 1 also shows that the DOS at E_F is pinned to its LDA value for $U \leq 3$ eV. This is the expected behavior for a Fermi liquid (FL) metal. With increasing U, however, our LDA+DMFT results show drastic modification of the spectral functions near E_F. Revealingly, in addition to large-scale SWT, we find that the FL-like pinning of the LDA+DMFT DOS to its LDA value, found for small U, is lost for $U = 4$ eV. Instead, the metallic state shows a clear pseudogap at E_F, with no Landau FL quasiparticles.

In Fig. 2, we compare our $U = 4$ eV (and, $U' = 2.6$ eV) results with PES for doped $FeSe_{1-x}$ [10]. Good semiquantitative agreement with experiment is visible for $n = 5.8$, where n is the total band filling of the iron d shell. In particular, the broad peak at ≈ -0.17 eV as well as the detailed form of the lineshape in PES is well reproduced by LDA+DMFT results for the hole doped case. This may suggest that the experiment could have been done on a tetragonal sample with small Selenium excess (we recall that exact stoichiometry is a sensitive issue in the FeSeTe alloys) [24]. For comparison, the computed LDA+DMFT spectra for the undoped ($n = 6.0$) and electron doped ($n = 6.1$) cases show progressively more disagreement with PES at low energies. However, the overall lineshapes, along with the peak around -0.2 eV and the low-energy pseudogap remain robust features in the DMFT calculation. In contrast to this, the correlated spectral functions close to E_F are insensitive to small changes in the electron (hole) concentration: we predict that combined PES/XAS on doped samples might show this in future.

We now focus on orbital resolved spectral functions of FeSe. Clear orbital-selective (OS)

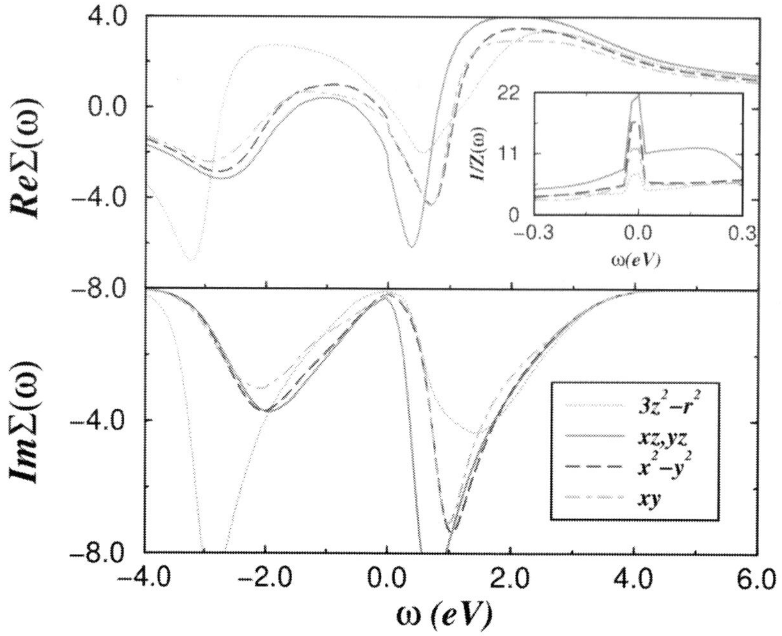

Figure 4. Orbital-resolved LDA+DMFT self-energies for electron-doped FeSe. Upper panel: Real parts showing a low-energy kink feature, at about 15 meV below E_F, in $\text{Re}\Sigma_a(\omega)$ with $a = xy, xz, yz, x^2 - y^2$. In the inset, we show the computed orbital-dependent effective masses. These are sizably enhanced relative to LDA values, in good quantitative accord with ARPES data [13]. Lower panel: The corresponding imaginary parts, showing clear sub-linear $(xy, xz, yz, x^2 - y^2)$ and almost quadratic $(3z^2 - r^2)$ in-ω dependence for $\omega \leq E_F$.

incoherence is visible in Fig. 3: a low-energy pseudogap is visible in the $xz, yz, x^2 - y^2$ DOS, and only the $xy, 3z^2 - r^2$ DOS show very narrow FL-like resonances at E_F. Examination of the self-energies in Fig. 4 shows that, for $n = 5.8$, only $\text{Im}\Sigma_{3z^2-r^2}(\omega) \simeq -a\omega^2$ for $\omega < E_F(= 0)$. Using the Kramers-Krönig relation, it follows that the Landau FL quasiparticle residue, Z vanishes near-identically for the $xz, yz, x^2 - y^2$ band carriers [from $\text{Re}\Sigma(E_F)$], direct numerical evaluation gives $Z_{xz,yz} = 0.046, Z_{x^2-y^2} = 0.059$). This translates into an effective mass enhancement $[\frac{m^*}{m} \equiv \frac{1}{Z} = 1 - \frac{d}{d\omega}Re\Sigma(\omega)|_{\omega=0}]$ of 21.5 for the xz, yz carriers and 17.0 for the $x^2 - y^2$ carriers, as shown in the inset of Fig. 4. This is in good accord with values estimated by an ARPES study on $FeSe_{0.42}Te_{0.58}$ superconductor [13], confirming the hypothesis about electronic correlations in FeSe made in that work. In our LDA+DMFT, these orbital-selective mass enhancements point toward the relevance of sizable MO electronic correlations in FeSe. However, we also notice that $d\Sigma/d\omega$ has appreciable frequency dependence at low energy: for a Landau FL metal, this quantity should be constant. Our finding of a frequency dependence in $d\Sigma/d\omega$ is thus fully consistent with a pseudogapped, incoherent metallic state as found above.

Finally, we shall point out that recent studies seem to be converging toward an intermediate-to-strong correlation scenario for the 122-Fe arsenides and chalcogenides [25, 26] as well as the 11-Fe selenides [27]. Semiquantitative agreement with the details of the PES lineshape along with specific description of transport [17] lends further credence to our view, which places the FeSe(Te) in the incoherent, bad-metallic regime of a sizably correlated MO Hubbard model. In earlier LDA+DMFT studies for the 1111-Fe pnictides [14] and 122-selenides [26], we found an incoherent metal normal state similar in many respects to the one shown here. Our study thus shows that sizable d-band electronic correlations are generic to the Fe-based superconductive materials.

3. Conclusion

To conclude, based on a five-orbital LDA+DMFT study, we have shown that orbital-selective incoherence characterizes the normal metallic phase in tetragonal FeSe. Good semiquantitative agreement with photoemission spectra and rationalization of a variety of unusual observations in a single picture lend support for our proposal. Sizable multi-orbital correlations are shown to be necessary to derive this orbital-selective incoherent metal. Emergence of superconductivity at low T, along with extreme sensitivity of the ground state(s) to minute perturbations in $FeSe_{1-x}Te_x$ compounds should thus be considered as manifestations of the myriad possible instabilities of such an incoherent non-Fermi liquid metal in close proximity to Mottness.

Acknowledgments

L.C.'s work was supported by CAPES - Proc. No. 002/2012. Acknowledgment is also made to FAPEMAT/CNPq (Projet: 685524/2010). S.L. acknowledges ZIH Dresden for computational time.

References

[1] Stewart G R 2011 *Rev. Mod. Phys.* **83** 1589
[2] Si Q and Abrahams E 2008 *Phys. Rev. Lett.* **101** 076401; Wu J, Phillips P and Castro Neto A H 2008 *Phys. Rev. Lett.* **101** 126401; Si Q *et al.* 2009 *New J. Phys.* **11**, 045001
[3] Imai T *et al.* 2009 *Phys. Rev. Lett.* **102** 177005
[4] Fang M H *et al.* 2008 *Phys. Rev. B* **78** 224503
[5] He S *et al.* 2013 *Nature Materials* **12** 605
[6] McQueen T M *et al.* 2009 *Phys. Rev. B* **79** 014522
[7] Garbarino G *et al.* 2009 *Europhys. Lett.* **86** 27001
[8] Masaki S *et al.* 2009 *J. Phys. Soc. Jpn.* **78** 063704; Miyoshi K *et al.* 2009 *J. Phys. Soc. Jpn.* **78** 093703
[9] Nie Y F *et al.* 2009 *Appl. Phys. Lett.* **94** 242505
[10] Yamasaki A *et al.* 2010 *Phys. Rev. B* **82** 184511; see also, arXiv:0902.3314.
[11] Yoshida R *et al.* 2009 *J. Phys. Soc. Jpn.* **78** 034708
[12] Subedi A *et al.* 2008 *Phys. Rev. B* **78** 134514; Lee K W, Pardo V and Pickett W E 2008 *Phys. Rev. B* **78** 174502
[13] Tamai A *et al.* 2010 *Phys. Rev. Lett.* **104** 097002
[14] Craco L *et al.* 2008 *Phys. Rev. B* **78** 134511; Laad M S *et al.* 2009 *Phys. Rev. B* **79** 024515
[15] Sales B C *et al.* 2009 *Phys. Rev. B* **79** 094521
[16] Williams A J *et al.* 2009 *J. Phys.: Condens. Matter* **21** 305701
[17] Craco L, Laad M S and Leoni S 2010 *Europhys. Lett.* **91** 27001; also, Craco L and Leoni S 2010 *Europhs. Lett.* **92** 67003
[18] Lee S H *et al.* 2010 *Phys. Rev. B* **81** 220502(R); Lee W C and Phillips P W 2012 *Phys. Rev. B* **86** 245113 Lee G *et al.* 2012 *Phys. Rev. Lett.* **109** 177001
[19] Kotliar G *et al.* 2006 *Rev. Mod. Phys.* **78** 865
[20] Hsu F C *et al.* 2008 *Proc. Natl. Acad. Sci.* **105** 14262
[21] Andersen O K 1975 *Phys. Rev. B* **12** 3060
[22] Craco L 2008 *Phys. Rev. B* **77** 125122
[23] Skornyakov S L *et al.* 2009 *Phys. Rev. B* **80** 092501
[24] Williams A J, McQueen T M and Cava R J 2009 *Sol. St. Comm.* **149** 1507
[25] Dagotto E 2013 *Rev. Mod. Phys.* **85** 849; Yu R and Si Q 2013 *Phys. Rev. Lett.* **110** 146402; Kutepov A *et al.* 2010 arXiv:1005.0885; Yi M *et al.* 2013 *Phys. Rev. Lett.* **110** 067003; Wang Z *et al.* 2013 arXiv:1309.6084
[26] Craco L, Laad M S and Leoni S 2011 *Phys. Rev. B* **84** 224520
[27] Liebsch A and Ishida H 2010 *Phys. Rev. B* **82** 155106; Aichhorn M *et al.* 2010 *Phys. Rev. B* **82** 064504

AUTHOR INDEX